DAVID GORLÆUS (1591-1612)

History of Science and Scholarship in the Netherlands, volume 13

The series *History of Science and Scholarship in the Netherlands* presents studies on a variety of subjects in the history of science, scholarship and academic institutions in the Netherlands.

Titles in this series

1. Rienk Vermij, *The Calvinist Copernicans. The reception of the new astronomy in the Dutch Republic, 1575-1750*, 2002, ISBN 90-6984-340-4
2. Gerhard Wiesenfeldt, *Leerer Raum in Minervas Haus. Experimentelle Naturlehre an der Universität Leiden, 1675-1715*, 2002, ISBN 90-6984-339-0
3. Rina Knoeff, *Herman Boerhaave (1668-1738). Calvinist chemist and physician*, 2002, ISBN 90-6984-342-0
4. Johanna Levelt Sengers, *How fluids unmix. Discoveries by the School of Van der Waals and Kamerlingh Onnes*, 2002, ISBN 90-6984-357-9
5. Jacques L.R. Touret and Robert P.W. Visser, editors, *Dutch pioneers of the earth sciences*, 2004, ISBN 90-6984-389-7
6. Renée E. Kistemaker, Natalya P. Kopaneva, Debora J. Meijers and Georgy Vilinbakhov, editors, *The Paper Museum of the Academy of Sciences in St Peterburg (c. 1725-1760), Introduction and Interpretation.* 2005, ISBN 90-6984-424-9, ISBN dvd 90-6984-425-7, ISBN book and dvd 90-6984-426-5
7. Charles van den Heuvel, *'De Huysbou.' A reconstruction of an unfinished treatise on architecture, town planning and civil engineering by Simon Stevin*, 2005, ISBN 90-6984-432-x
8. Florike Egmond, Paul Hoftijzer and Robert P.W. Visser, editors, *Carolus Clusius. Towards a cultural history of a Renaissance naturalist*, 2007, ISBN 978-90-6984-506-7
9. Lissa Roberts, Simon Schaffer, Peter Dear, editors, *The mindful hand: inquiry and invention from the late Renaissance to early industrialization*, 2007, ISBN 978-90-6984-483-1
10. Dirk van Delft, *Freezing physics. Heike Kamerlingh Onnes and the quest for cold*, 2007, ISBN 978-90-6984-519-7
11. Patricia E. Faasse, *In splendid isolation. A history of the Willie Commelin Scholten Phytopathology Laboratory 1894-1992*, 2008, ISBN 978-90-6984-541-8
12. Albert van Helden, Sven Dupré, Rob van Gent & Huib Zuidervaart, editors, *The origins of the telescope*, 2010, ISBN 978-90-6984-615-6
13. Christoph Lüthy, *David Gorlaeus (1591-1612). An Enigmatic Figure in the History of Philosophy and Science*, 2012, ISBN 978-90-8964-438-1

David Gorlæus (1591-1612)

An Enigmatic Figure in the History of Philosophy and Science

Christoph Lüthy

Amsterdam University Press

The publication of this book is made possible by a grant from the J.E. Jurriaanse Foundation, the Evert Willem Beth Foundation, the Foundation Sormani Fund, the M.A.O.C. Gravin van Bylandt Foundation, the Douwe Kalma Stifting, and the OAPEN Foundation.

This book is published in print and online through the online OAPEN library (www.oapen.org). OAPEN (Open Access Publishing in European Networks) is a collaborative initiative to develop and implement a sustainable Open Access publication model for academic books in the Humanities and Social Sciences. The OAPEN Library aims to improve the visibility and usability of high quality academic research by aggregating peer reviewed Open Access publications from across Europe.

Cover illustration: Head of a young man. Detail from the tomb of David Gorlaeus in Cornjum (see page 13, fig. 1) (Photograph by Klaas Tijdsma)
Cover design: Kok Korpershoek, Amsterdam
Lay-out: Heymans & Vanhove, Goes

ISBN 978 90 8964 438 1
e-ISBN 978 90 4851 680 3 (pdf)
e-ISBN 978 90 4851 681 0 (ePub)
NUR 680/685

Table of Contents

Preface

I came across the name of David Gorlaeus for the first time while working on my doctoral dissertation on seventeenth-century matter theory in the early 1990s. The dazzling diversity of the authors who pleaded for the existence of atoms in the period 1590-1630 puzzled me greatly as I could find neither a coherent pattern nor an overarching concern in the various antiquarian, historical, theological, metaphysical, physical, alchemical and microscopic reasons that they offered. In a number of publications I have since examined a range of particularly puzzling figures or types of argumentation in favor of the existence of atoms. When life's circumstances took me to the Netherlands, where I have settled, I could not avoid turning my attention to David Gorlaeus, who seemed to me a particularly elusive figure. After all, very little was known about his life, and his ideas were particularly hard to place as they mixed metaphysics and natural philosophy in a markedly unusual way and in unexpected moments added observations taken from the fields of astronomy, optics and chemistry. When I read that the author had passed away at age 21, and that he was moreover starting out as a theology student and was not a person engaged in empirical research, my initial curiosity increased even further, turning into a detective's quest for the reconstruction of the circumstances that led to an inexplicable fact. The more I searched, the more I became convinced that Gorlaeus was an unusually talented thinker of extraordinary originality and maturity, notably when one considered the young age at which he wrote his works. In fact, I remain persuaded that his philosophical synthesis renders him one of the early seventeenth century's most brilliant Dutch intellects. Had he been granted more years to live and the chance to develop his thoughts further, so I now imagine, he might well have become as radical and famous a thinker as Spinoza. Although such counterfactual musings do not belong to the historian's task, they do in this particular case explain one of the main emotional reasons for investigating the short life of this talented thinker.

Some of my findings concerning Gorlaeus have been published before, but in places that are not easily accessible. When it was decided to organize an academic celebration on the occasion of the 400th anniversary of Gorlaeus' death, to be held in Cornjum on 27 April 2012, it was pointed out to me that in the absence of any monographic study of this thinker, it would be difficult to persuade the larger public of the status that I wished to claim for my young thinker. This justified observation has led me to write down everything I have so far managed to uncover about the life, circumstances and thoughts of David Gorlaeus. I realize of course that much remains to be found, in manuscript and published sources, about his family, personal circumstances and impact on the evolution of seventeenth-century philosophy and science. For this reason, the present book cannot offer more than a provisional account and is written in the hope that the story told here will inspire other historians of philosophy, science and theology to take over where I have left off.

It gives me great pleasure to thank a number of persons who have helped me in this enterprise. The long section on Gorlaeus' teacher Henricus de Veno has benefitted enormously from the contribution of Leen Spruit, who found De Veno's inquisitorial acts in the Vatican. Arjen Dijkstra has joined me in the hunt for Gorlaeus' Frisian circumstances; a profound expert on early modern Friesland, he has uncovered a number of important facts, texts, and connections that I am pleased to acknowledge in the relevant passages. Gerben Wierda, a formidable archival hunter, has contributed considerably to my attempts to reconstruct Gorlaeus' family circumstances. Once again, whatever I owe to him is gratefully acknowledged in the footnotes. My research on Gorlaeus has also benefitted from the expert advice of Sander de Boer, Theo Bögels, Erik-Jan Bos, Jos van den Broek, Robin Buning, Davide Cellamare, Paul Dijstelberge, Martin Engels, Paul Hoftijzer, Ulrich G. Leinsle, Ferenc Postma, Jarich Renema, Jacob van Sluis, Jaap van der Veen and Huib Zuidervaart, all of whom I would like to thank most emphatically.

I am particularly grateful to four extremely knowledgeable experts in the history of Dutch philosophy and science, who have carefully examined the final draft of this book: Klaas van Berkel, Theo Verbeek, Han van Ruler and Chungling Kwa have contributed to a substantial improvement of my account. Amsterdam University Press, and notably Anniek Meinders and Maaike Groot, deserve to be praised here for the competent way in which they accompanied me in the production of this book. I should furthermore like to thank Thomas Swann for his excellent editorial work. I am also grateful to Brill Academic Publishers and the *Renaissance Quarterly* for allowing me to reproduce passages from earlier publications on Gorlaeus and De Veno.

Finally, I must thank the two organizations that have sponsored my research. Ten years ago, the Royal Netherlands Academy of Arts and Sciences (KNAW) provided me with a luxurious fellowship that allowed me to carry out my initial investigations into Gorlaeus. Thereafter, I was able to conduct most of my research within a programme sponsored by the Netherlands Organisation for Scientific

Research (NWO). In these times of shrinking research budgets, such generous donors deserve to be mentioned with particular gratitude.

I dedicate this book to Carla Rita Palmerino, with whom I enthusiastically share a home, a university office and thus my entire life, and to our two boys, Tommaso and Filippo, who are growing up far too quickly.

Chapter 1

Introducing Gorlæus

Human history presents itself in stories about the past. Wherever these stories are not records of first-hand experience, they are based on the examination of archival materials, old books, archeological digs, paintings or material objects. History books present smaller or larger segments of this past in an organized narrative and inevitably from a certain perspective. Taken together, these books provide something like a large map of everything of which we know or believe, on the basis of a rational reconstruction, that it has been the case or has happened.

The resulting composite map of the past might be compared to those early nineteenth-century maps of Africa, on which strategically important elements, such as coastal lines, islands, estuaries and the main rivers and their larger tributaries, are drawn with great accuracy; while other, less accessible areas are indicated as blanks, as unstructured pieces of terra incognita. Like the explorer, the historian feels the allure of those unknown regions, and particularly of those regions that seem to hide a mystery or hold out a particular promise.

The present book is the result of such a blank spot and the presence of a particularly alluring mystery. The mystery to be explored carries the name of David van Goorle, an early modern Dutchman who is better known as Gorlaeus, his Latinate name, which we shall use in this book.

The reason why it seems desirable to reconstruct and narrate Gorlaeus' life, thought and influence is threefold. The first reason has to do with the fact that Gorlaeus died at the mere age of 21, but left behind two manuscripts, published posthumously, that testify to an extraordinary intellectual maturity. *Wunderkinder* are usually found in music or poetry, but not in systematic philosophy. The attempt undertaken in this book to capture as much as possible of his life and intellectual circumstances is the result of the desire to understand the author of this premature work as well as the forces that led him to write such books in his late teens. The second reason is that, irrespective of Gorlaeus' precocious age, his philosophical and scientific thought is unusual, fascinating and in several respects ahead of its time. Indeed, until about 1650, he was regarded as one the most important European innovators (*novatores*) in philosophy. Yet, despite his early fame, Gorlaeus has not yet found a secure place in the historiography of early modern

Dutch intellectual life. The third reason for dedicating a book to him is that Gorlaeus' philosophical and scientific proposals appear to have exercised a notable influence on the evolution of Dutch thought and most interestingly on the ideas of Descartes' early associates.

Gorlaeus' thoughts and circumstances have never been investigated in a comprehensive way. This study attempts to fill this lacuna. However, it is evident to the present explorer that additional expeditions will be required to map this territory completely. The reader is therefore asked to regard the present book as a provisional expedition report and as a call for further explorations.

1.1. THE TOMB

David Gorlaeus died on 27 April 1612, at the young age of 21. He was buried in the village church of Cornjum, in the Dutch province of Friesland, a few footsteps from the aristocratic mansion in which his parents resided. Both the location and the elegance of his tomb reflect the elevated status of the family to which he belonged. Today's visitor can visit his grave, which lies under a glass pane right in the middle of Cornjum's handsome church, below the pulpit and visible from all the pews that line the church's four walls. When the black carpet that usually covers it is removed, the onlooker must in fact fight the sensation that the church has been purposefully built around Gorlaeus' centrally placed tomb.

Like a ribbon along the rim of the tombstone (see Figure 1) runs a Dutch sentence that provides some factual bits of information: "In the year 1612, on 27 April, the very erudite and wise youth David van Goorle died, who is buried here."[1] The middle of the tombstone features a coat of arms (which was brutally disfigured during the French occupation in 1796), together with a Latin poem, "To the honor and memory of the splendid youth David Gorlaeus." Its iambic hexameters translate as follows:

> Here lies buried that flower of youth,
> Gorlaeus, taken away in the very spring of his life,
> While he was rising to the highest endeavor of ancient praise.
> Death, which does not allow anything sublime to last for long,
> Has carried away from the Earth this ornament, which is due to the Heavens,
> As the Earth was incapable of carrying such a gift.
> This very illustrious mind and heavenly spirit,
> Freed from its fetters and the weight of the body,
> Sought the Havens, whence had come its seed.
> And there it [*sc.* Gorlaeus' spirit] views Him who is born through the Eternal Father,

FIG. 1: Gorlaeus' tomb in the church of Cornjum. Today, it lies under a glass pane which is inserted in the wooden floor of the church. (Photo: Klaas Tijdsma)

Who through His death has atoned for the human crime,
Christ, the world's mediator and giver of peace.
This is safety; this is the peak of the highest good,
Greater than which our mind cannot desire anything.[2]

The task of epitaphs is to be excessive in lament and flattery alike. This particular tombstone might however be accused of excessive restraint in both Dutch and Latin. Gorlaeus was not just a promising youth whose life was broken before it reached its bloom. Judging by his writings, which were to appear in print a number of years after his death, he may well have been one of the most original thinkers of early modern Dutch intellectual history. Eight years after his death, in 1620, a first, densely argued treatise appeared under his name, which carried the following long title: "Philosophical Exercises (*Exercitationes philosophicae*), in which the entire body of theoretical philosophy is discussed, and in which several essential dogmas of the Aristotelians are overturned."[3] Fully three decades later, in 1651, a second and much shorter treatise saw the light of the day, entitled *Idea physicae* ("Sketch of Physics").[4] This book is conspicuous for the brevity, precision and boldness with which it attempted to cast a new basis for physics.

1.2. GORLAEUS IN THE HISTORIOGRAPHY OF PHILOSOPHY

Because of his early death, Gorlaeus' fame among his contemporaries rested almost exclusively on their acquaintance with the contents of the *Exercitationes* and to a much lesser degree of the *Idea*, which was published late and enjoyed a much more limited circulation. Even in the Netherlands, only few readers knew who their author was or what had motivated him. The extended 1643 edition of Valerius Andreas' *Bibliotheca Belgica*, for example, only contains the following barren and uninformative entry:

> David Gorlaeus, from Utrecht, published with the types of Commelinus his *Philosophical Exercises, in which the Entire Body of Theoretical Philosophy Is Discussed*, 1620, 8°.[5]

Jean-François Foppens' bibliographical encyclopedia of 1739, identically named *Bibliotheca Belgica*, provides in its 1168 pages a plethora of information about the lives and works of all known authors from the Low Countries. And yet, unaware even of Gorlaeus' *Idea physica,* which had been published in the meantime, it simply reiterates Andreas' laconic entry of a century earlier.[6]

For the general public, Gorlaeus' intellectual reputation was entirely disembodied: there existed two treatises proposing a series of uncommon ideas, and their title pages sported their author's name, yet there was no historical personage to whom one could have attached these ideas, nor a *vita* that could have shed light on them. As a possibly inevitable consequence, Gorlaeus ended up entering the history books under a number of different guises. Fictitious identities had to stand in as the lieutenants of an absent historical persona.

We encounter the first of his multiple personalities merely two hundred steps from the church of Cornjum, where Gorlaeus lies buried. There, the visitor comes across a burial mound covered by beautiful old trees and containing the tombs of the last inhabitants of Martenastate, the mansion in which Gorlaeus' parents had lived. A signboard placed there so as to explain the site to the passer-by refers to David Gorlaeus as "the Dutch Galileo." The board fails to motivate this comparison, although it is probably due to a parenthetical remark in Dijksterhuis' *Mechanisation of the World Picture*. There, Gorlaeus is said to have anticipated Galileo's distinction between the geometric-mechanistic properties of matter and those secondary, sensory properties that are generated merely in the perceiver.[7] However, not only has Dijksterhuis here misread Gorlaeus, but the comparison between Galileo and Gorlaeus is also generally misleading. But then, one wonders, which comparison wouldn't be? Both as an historical figure and as a thinker, Gorlaeus is hard to label and to compare. The anomaly of his case begins of course with his early death, which implied that he passed away before he had acquired any fame, quite unlike the world-renowned Italian with whom the signboard compares him. Secondly, whereas it is easy to explain Galileo's fame, for example, by reference to his telescopic discoveries, his law of free fall or the heliocentric views for which he was condemned by the Inquisition, Gorlaeus' achievements are decidedly more difficult to explain in a few words to the tourist visiting Cornjum.

But what did seventeenth-century readers think of his work? The first reactions stem from the 1620s, soon after the *Exercitationes* had been published. In his belligerent youth, the future irenic 'Secretary of the Republic of Letters', Marin Mersenne, repeatedly invoked Gorlaeus' name among a host of important anti-Aristotelians and heretics who needed combating: together with Patrizi, Basson, Bodin, Carpenter, Hill and Olivi, Gorlaeus is bashed for his anti-Aristotelianism. Their attitude of opposition was both futile and arrogant, Mersenne wrote, because "Aristotle is an eagle in philosophy, while these others are mere chicks, who wish to fly even before they have any wings."[8] In 1624, Mersenne announced his intention to publish an encyclopedia "in favor of all truths against all kinds of lies, in which I shall carefully examine the views advanced by Gorlaeus, Carpenter, Basson, Hill, Campanella, Bruno, Vanini, and some others."[9] Such was thus the hostility with which Gorlaeus' *Exercitationes* were greeted by those who defended the inherited philosophy. But such were also the names with which he was associated: at least Giordano Bruno (1548-1600), Tommaso Campanella (1568-1639) and Jean Bodin (1630-1596) still feature prominently in the historiography of philosophy.

Although Mersenne himself would soon thereafter transform himself into a model of open-mindedness, those who remained faithful to the traditional ways of teaching philosophy felt similarly opposed to Gorlaeus in later decades. In 1641, for example, the Utrecht theologian Gijsbert Voetius described Gorlaeus as a thinker who had committed theologically dangerous fallacies, "misled by his youth."[10] And again twenty years later, in 1662, the influential Franeker professor of philosophy Arnold Verhel deplored the miserable conditions of contemporary metaphysics: "I do not understand what fatal catastrophe has overcome our philosophical studies in this deplorable age." Metaphysics, he cried, was everywhere under siege, and its enemies were seen to triumph at the universities, on the pulpit, in the courtroom.

> Moreover, in their noisy brawls and jeers they revile Aristotle himself, the father of metaphysics and the prince of all philosophers. Against him grunts the zeal of the Ramists, the gainsaying of the Gorlaeans, the high-browed arrogance of the Cartesians, and the authority of certain teaching doctors.[11]

This reference to 'Gorlaeans' is conspicuous, as it suggests that there existed a current of philosophers who worked explicitly in Gorlaeus' tradition.

Towards the end of the century, however, there were few north-European philosophers left to defend Aristotle with drawn sword. The majority view was that the Aristotelian system needed serious overhaul if not total replacement. The battle being over and won, anti-Aristotelianism came to resemble the proverbial beating of the dead horse, as historians turned to writing the pre-history of this victory over the scholastic system. As must be evident, Gorlaeus plays a positive role in these narrations. Take, for example, the historian of philosophy Daniel Georg Morhof (1639-1691), to whom Gorlaeus' *Exercitationes* seemed

> quite ingenious. It opens with a treatment of philosophy in general and then moves on to metaphysics and logic, stating many things that deviate from the Aristotelian view. It subsequently turns to physics, in which it pursues its own hypotheses, attacking Aristotle's. The principal hypotheses are that the heavens are nothing else but the extension of air; it also postulates only two elements of mixture, namely earth and water, for it excludes fire from the list of elements, defining it as a mere accident. This was also the dogma of the first Cartesians.[12]

Mersenne, Voetius, Verhel and Morhof document that the most obvious label that contemporaries stuck to Gorlaeus was that of anti-Aristotelianism. In fact, Burman's eighteenth-century *Traiectus eruditus* opens its entry on Gorlaeus with the words: "He was a famous author, and is counted among those who dared to battle against Aristotle."[13]

But Morhof's short characterization of Gorlaeus' work, which we have just cited, also comments on the order in which the philosophical disciplines are presented. The *Exercitationes* starts, so Morhof tells us, with a definition of philosophy in general, subsequently casts a metaphysical basis, turns to logic and finally reaches physics, a domain in which it comes up with new results. Noteworthy about this description is the suggestion that Gorlaeus anticipated "the first Cartesians" in certain respects. These two themes – the order in which Gorlaeus presents his philosophical doctrines and his possible influence on Descartes and the Cartesians – frequently return in early modern comments on Gorlaeus. In several of them, Gorlaeus is said to have anticipated not only a number of Cartesian doctrines, but also certain features of the structure of Descartes' system. To understand better how this view might have arisen, let us cite a passage from Daniel Garber's pathbreaking study, *Descartes' Metaphysical Physics*:

> It is fair to say that [Descartes'] view of the order of knowledge may well have presented a significant departure from the mainstream of the scholastic tradition.

> Though there were many differences between different scholastic writers, there was wide agreement that knowledge of physics is largely independent of knowledge of metaphysics, however precisely either discipline is defined. And so, they claimed, one can (and, in fact, ought to) study physics before undertaking the more elevated studies of God and being as such that pertain to first philosophy. In demanding that physics must be grounded in some sense in metaphysics, in knowledge of God and the soul, Descartes is stepping clearly outside that tradition. And so when by the 1630s Descartes came to hold to the priority of metaphysics in the strong sense he held it, his view would likely have been recognized as a clear departure from the received view.[14]

But it seems to have been exactly this 'departure' that some early modern readers claimed that Gorlaeus had carried out before Descartes. For indeed, as Morhof pointed out and as will be shown in detail in chapter 2, Gorlaeus' physics is, at least in the *Exercitationes*, made to depend on his metaphysics.

But in addition to this possible methodological anticipation, early modern historians of philosophy also commented on a real moment of contact between Cartesianism and Gorlaeus. When discussing the latter's philosophy in his famous *Dictionary*, Pierre Bayle reported the following incident:

> When Regius, a disciple of Descartes, was harassed for a thesis concerning the union of the soul with the body, he claimed that he had merely used Gorlaeus' own terms. That did not however help him in the least; as a consequence, Voetius, professor of theology, flung as much dirt at Gorlaeus' views as he could.[15]

Considering this specific episode as well as certain methodological and doctrinal overlaps between Gorlaeus and Descartes, some early modern commentators jumped to far-reaching conclusions. Morhof felt that Gorlaeus "certainly deserves praise for having recognized before Descartes what Descartes later wanted to make appear as his own doctrines."[16] These words amount to the claim that Descartes took some of Gorlaeus' methods and doctrines and sold them as his own. Jacob Friedrich Reimmann, another early modern German historian of philosophy, made the additional claim that after the 1641 clash between Regius and Voetius, "Cartesians accepted most Gorlaean theses into their system," suggesting that Gorlaeus' philosophy was to become an integral and constitutive part of Cartesianism.[17]

These surprising claims deserve to be examined with care. That Descartes was indebted to the Dutchman Isaac Beeckman is well known.[18] By contrast, there is no modern scholarship that has taken the claims of Morhof or Reimmann seriously or even examined them. Such an examination will be carried out in our fourth chapter.

1.3. GORLAEUS IN THE HISTORIOGRAPHY OF SCIENCE

With respect to the historiography of philosophy, the situation presents itself roughly as follows: while early modern historiographers appreciated, and possibly lionized, Gorlaeus' role as an anti-Aristotelian reformer of philosophy and as a possible ancestor of Cartesianism, most contemporary historians of philosophy have pretty much forgotten him. Precisely the opposite development has occurred in the history of science. There, Gorlaeus has in the past 120 years come to play a completely different and in fact more prominent role, namely as a pioneer of atomism. He acquired this label, and the fame that has come with it, in the late nineteenth century; that is, in the period that witnessed renewed debates in physics, chemistry and philosophy over the existence of atoms.

Two things ought be remembered about the label 'atomism', however. The first is that this term is an early modern neologism. According to Robert and Henri Estienne's *Thesaurus Linguae Graecae* of 1572, it first appeared as a Greek term in the *Antirrheticon* (ca. 1470) of Theodor Gazes (1400-1475). In his *Democritus reviviscens* of 1644, Jean-Chrysostôme Magnen spoke of a 'Philosophy of Atoms'. But the term 'atomism' only appeared in the last quarter of the seventeenth century, possibly first in Ralph Cudworth, who used the term 'atomicism' to denote a materialist and atheist version of a true corpuscular philosophy.[19] The second point to keep in mind is that early modern authors did not use 'atomism' when grouping philosophical positions into camps. Robert Pasnau's allegation that "atomism [was] a view that barely mattered," is correct, "insomuch as very little turns on whether one thinks the material realm is or is not infinitely divisible."[20] It was not the metaphysical question of divisibility that ultimately defined the camps. Take Descartes, who was an 'anti-atomist' and yet provided the first illustrated guide to Democritus' materialistic world of particle-filled vortices; or the late Daniel Sennert of the *Hypomnemata physica*, who could be viewed as an 'atomist' simply because he postulated the existence of such physical indivisibles, but who took his atoms to be the carriers of the substantial forms that Descartes so adamantly rejected. As a consequence of his particular physics, Descartes would routinely be grouped with the Epicureans (to his great displeasure, of course); while Sennert was often grouped with the Aristotelians.

Historically speaking, the doctrine that all matter is ultimately composed of indivisible particles is of course a doctrine of ancient Greek extraction. But because of Aristotle's numerous intelligent objections to Democritus' physics, the scholastic tradition had presented the idea of indivisible yet extended magnitudes as a schoolbook example of erroneous reasoning. Only in the later sixteenth and seventeenth centuries did an atomic view of matter manage to resurface in any significant way. The reasons for this atomistic revival will be discussed later. What is of importance for our present purposes is that, partly in response to Gustav Theodor Fechner's Neo-Kantian approach to atomism in *Über die physikalische und philosophische Atomenlehre* of 1855, a 'historico-critical method' was developed that sought epistemological answers to the atomistic debate by examining the historical sources. A typical expression of this approach is found in Arthur Hannequin's *Essai critique sur l'hypothèse des atomes* of 1895, where we read: "The contemporary theories are thus in agreement with history on this point: they give their blessing to the predominance of the atomist hypothesis."[21] The most acute and reliable among these historico-critical authors was Kurd Lasswitz, whose admirable two-volume *Geschichte der Atomistik* still constitutes an historiographic treasure trove.[22] It was Lasswitz who rediscovered Gorlaeus and labelled him an 'atomist'. Not least because of the fact that he also had access to one of the exceptionally rare copies of the *Idea physicae*, his analysis of Gorlaeus' philosophical and physical system went beyond everything that had previously been written about this author. Lasswitz described the ontological basis of Gorlaeus' matter theory, its link to a nominalist logic, its theory of substances and qualities and the resulting atomistic physics.

The path by which Gorlaeus arrived at his atoms, however, looked unlike anything that Lasswitz had encountered elsewhere. He therefore tried to obtain information about this mysterious author. The bits of information that the Dutch scholars he contacted could provide him with seemed inconclusive. The most important discovery was that Gorlaeus had enrolled as a student in theology in Leiden in 1611 – a disciplinary background to atomism that Lasswitz had not expected. Finding that the title of the *Exercitationes* of 1620 merely mentioned that the author had in the meantime passed away, and considering that most other early modern atomists published their views after 1620, Lasswitz remained puzzled. Unable to place the author and his treatises intellectually, he made the following appeal: "A monograph on Gorlaeus and this important decade is a great desideratum."[23]

Thanks to Lasswitz, Gorlaeus entered the historiography of atomism in 1890 and he has been treated as an atomist ever since. As such, he came to enjoy a new career as a scientific author and matter theorist. Even historians of philosophy now came to regard him in this specific light. The Dutch historian of philosophy Jan Pieter Nicolaas Land, while praising Gorlaeus' scientific curiosity, depicted him at the same time as a victim of his theological training: "Had he paid more attention to the natural phenomena and the principles of motion, his project would have been crowned with greater success than his particular education for the church seems to have allowed for."[24] For Land, then, Gorlaeus' strivings were essentially 'scientific', but his disciplinary choice for theology revealed itself as an intellectual impediment. With a greater sense of the historically plausible – after all, the laws of motion were defined only after Gorlaeus had already passed away – Ferdinand

Sassen characterized Gorlaeus as a "lonely figure" who "liberated himself in important respects from the Aristotelian physics" long before others, "attempting to replace it with an atomistic natural philosophy."[25] In one word, then, after Lasswitz, Gorlaeus has generally been perceived as a proto-scientist, even though historians found it difficult to establish the value of his scientific contribution. "Maybe a qualified author will one day find the time to assess the merits and shortcomings of a man who did his best to become one of the reformers of science," Land suggests, shrugging off his perplexity.[26]

Gorlaeus has been unable to shake off his taxonomic species name ever since, being remembered either as an 'atomist' or not at all. Frans Maurits Jaeger's important study of 1918, which to this day constitutes the most fruitful attempt to fill the historiographic lacuna spotted by Lasswitz, carries the title "On David van Goorle as an Atomist."[27] Andreas van Melsen's *From Atomos to Atom* repeatedly turns its attention to the doctrines of "the Dutch atomist David van Goorle."[28] Tullio Gregory's study, which bundles up Gorlaeus with the chemist and Wittenberg professor of medicine Daniel Sennert, appeared as a part of his "Studies on Seventeenth-Century Atomism."[29] In Dijksterhuis' *Mechanisation of the World Picture*, Gorlaeus appears as a precursor of the atomistic treatment of qualities in Galileo's *Assayer*; Hooykaas mentions him in one breath with Isaac Beeckman, considering these two men of "Flemish-Calvinist" extraction as "being among the first who based their physical explanations entirely on the atomistic doctrine."[30]

For twentieth-century historians of science, the importance of Gorlaeus lies exclusively in the fact that he was among the first to have rejected a scholastic theory of matter and its substantial forms in favor of an atomistic theory, stating that "nothing is real in bodies apart from the atoms."[31] His insistence, as Jaeger puts it, that "no rational explanation of natural phenomena is possible without the acceptance of the idea of an atomistic structure of matter," made him appear as a pioneer in the history of a concept that had begun as a metaphysical proposal in Greek antiquity, had been suppressed for centuries, had begun to resurface in the late sixteenth century and would win its ultimate triumph around 1900.[32]

In the process, Gorlaeus' fame became utterly lopsided: from the anti-Aristotelian philosophical *novator*, as seventeenth-century readers had understood him, he had turned into an empirical natural scientist. Ignoring its metaphysical anchoring, for example, the eminent historian of chemistry James Partington praised Gorlaeus' "scientific" form of atomism, which he contrasted with the philosophical "speculations" of Giordano Bruno; the latter he considered to be "of no physical importance."[33] In his historical survey, *Elements and Atoms Once and Now*, Jaeger confidently described Gorlaeus as a predecessor of Jan Baptist Van Helmont and Robert Boyle because of this quantitative and structural approach to matter.[34]

A new step in Gorlaeus' transformation into a scientist is reached in Lancelot

Law Whyte's *Essay on Atomism*, where our young hero, misspelled as "Garlaeus," appears in a list of fourteen important men who coupled a "new attention to quantity [...] with the lively interest in atomism." The other names are: Bodin, Galileo, Basson, Sennert, Berigard, Borelli, Huygens, Magnen, Charleton, Gassendi, Boyle, Leeuwenhoek, Newton and Halley.[35] Whyte has appended Gorlaeus' portrait in an imposing gallery. But does Gorlaeus really belong to this family? He only does if one recognizes the incongruity of this 'family' and applies to it Wittgenstein's notion of partial family resemblances. At the same time, however, this insertion into the pedigree of 'scientific atomism' does violence to both his larger project and to his particular intentions.

The apotheosis of this scientific *persona* is to be found in Leiden University's decision to name its largest science laboratory as well as its science library after Gorlaeus (see Figure 2). Incidentally, how the *Gorlaeus Laboratory* received its name deserves to be mentioned here, because it says much about the perils to which forgotten authors are exposed. Egbert Havinga, a professor of chemistry who had overseen the construction of the new chemistry facilities, had in truth proposed a different name. If it had been up to him, the buildings would have been named *Sylvius Laboratory* – after Franciscus de la Boë, called Sylvius in Latin (1614-1672). In 1669, Sylvius had effectively set up the Netherlands' first chemical university

FIG. 2: The Gorlaeus Laboratories at Leiden University. (Courtesy of Gorlaeus Laboratories)

laboratory at Leiden University. However, Professor Havinga's proposal met with fierce opposition from the students, who by 1970 had conquered the right to be involved in the running of the university's faculties. A student called Frans van Kleef went to the University Archives to check up on Sylvius. What he found was subsequently printed in *Chimica*, the university's chemistry journal. Sylvius was no chemist, Van Kleef protested, but a physician (which of course is true, as his chair was in medicine, like that of many other early modern chemists); he had furthermore fathered an illegitimate child (which was an odd complaint coming from the lips of a rebellious student); and, finally, students were overheard punning about the *Syphilislab*. The conclusion was damning indeed:

> With their choice of a name for the chemical building complex, the present subfaculty has made itself guilty of laziness, of the entirely improvised proposal of a name without any further desire to get to know anything about the man behind that name, and subsequently of an attempt to falsify history. Thrice shame on them![36]

Professor Havinga had to withdraw his proposal. The faculty journal *Chimica* celebrated this as "a proof of the fact that students really do have a right of participation in faculty matters."[37] 'Gorlaeus' became the new proposal. A student called Reinoud commented in *Chimica*: "I do not know who Gorlaeus was or is, but that shall probably be investigated in the near future."[38] Had the research been properly conducted, there should of course have been plenty of reason to reject Gorlaeus and return to the original proposal; but the two authors, who based themselves (among other things) on Partington and Van Nieuwenburg's *Short History of Chemistry*, and cited Gorlaeus' longer work wrongly as *Exercitationes physicae*, somehow managed to convince themselves that Gorlaeus was an appropriate namesake for the new laboratories.[39] Merely out of a sense of historical justice, it ought to be added here that nowadays Leiden also boasts a Sylvius Laboratory, suitably situated on Sylvius Street.

But we must return to Gorlaeus himself. Irrespective of whether his atomism did influence the subsequent evolution of an atomistic conception of matter – that such an influence existed will first have to be documented – his depiction as a pioneering natural scientist is clearly excessive for three reasons. First, his two extant treatises nowhere proceed along empirical lines. While it is true, as we shall see, that his works contain the occasional reference to optical, astronomical and chemical observations, his argumentation is rooted in metaphysics and natural philosophy. Second, Gorlaeus' short biography simply does not allow for much experimental practise. When he died at age 21, he was a first-year theology student who presumably had an Arts degree from Franeker University in his pocket. While his young age and his chosen discipline do not rule out an interest in the nascent experimental sciences, there is no circumstantial evidence to allow for the conclusion that his theory of matter was driven by first-hand experimental evidence, let alone by chemical practise.

In order to understand Gorlaeus' project, one must therefore find an answer to the following question: what may have brought a twenty-year old student to develop a new philosophy, and one moreover that relies on the existence of atoms? It is one of the chief objectives of this book to provide an answer to this question.

Chapter 2

Gorlæus' Two Treatises

In order to be able to appreciate Gorlaeus' place in the intellectual landscape of the opening years of the seventeenth century, and to assess his possible contribution to the history of philosophy and science, we must first acquaint ourselves with his ideas. Since his works are in Latin, and no one has ever provided a detailed synthesis of their contents, it is inevitable that we ignore the traditional order of presentation and begin with a synopsis of his thought before reconstructing his life.

2.1. METHOD OF PRESENTATION

There are different ways in which one can expound the ideas of a philosopher. The most obvious manner of presenting Gorlaeus' philosophy is by condensing the arguments of his two works into a single paraphrase. This is because the *Exercitationes philosophicae* (printed in 1620) and the *Idea physicae* (printed in 1651) bear a straightforward relation to each other. The 352-page *Exercitationes* tries to anchor Gorlaeus' natural philosophy in an ontology, or philosophy of being. By contrast, the scope of the 76-page short *Idea* is more straightforward and simple: it limits itself to the domain of natural philosophy (*physica*) and anticipates the doctrines of natural philosophy of the latter parts of the *Exercitationes*. The relation between the two books can be gauged by a comparison of their thematic structure. The *Idea physicae* is composed of thirteen chapters, which rehearse the traditional succession of themes in courses of natural philosophy:

Ch. 1: Which treats of the constitution of physics and nature
Ch. 2: On (what are commonly called) the internal principles of nature
Ch. 3: On the external principles of natural things
Ch. 4: On composition, quantum and the continuum
Ch. 5: On motion, place and time
Ch. 6: On the heavens
Ch. 7: On the elements and mixture
Ch. 8: On meteors

Ch. 9: On metals, the soul, life and death
Ch. 10: On the vegetative soul
Ch. 11: On the sentient soul
Ch. 12: On the main qualities affecting the senses
Ch. 13: On the human soul[1]

Although in his *Idea physicae* Gorlaeus redefines a host of Aristotelian doctrines, its structure nevertheless mirrors that of traditional textbooks; these in turn follow the canonical order of the Aristotelian works: from the *Physics* by way of *De caelo*, *De generatione et corruptione* and *Meteorology* to the treatise *De anima*. The *Exercitationes philosophicae*, by contrast, has a more innovative structure, although the traditional order remains visible in its eighteen *exercises*, which carry the following titles:

Ex. I: (No title) [Definition and function of philosophy; refutation of traditional metaphysics]
Ex. II: On being
Ex. III: On distinctions
Ex. IV: On the universal and singular
Ex. V: On the accident
Ex. VI: On quantity
Ex. VII: On quality
Ex. VIII: On relations
Ex. IX: On motion
Ex. X: On place
Ex. XI: On time
Ex. XII: On the composite
Ex. XIII: On atoms
Ex. XIV: On matter and form
Ex. XV: On the coming-about and perishing of things
Ex. XVI: On the heavens
Ex. XVII: (No title) [On the elements; and that the Earth does not move]
Ex. XVIII: On the soul[2]

This succession of themes represents a noteworthy and novel blend of a physical succession (comparable to that found in the *Idea physicae*) with a different, logico-methodological order. It starts from a definition of philosophy and proceeds by way of an ontological definition of being and a characterization of mental concepts to the categories of being and their composition. Let us anticipate that this order of presentation is half-way between the traditional structure of traditional textbooks in natural philosophy and that of Descartes' *Principles of Philosophy*.

I have not found any substantial doctrinal differences that would permit me to define the chronological relation of the *Idea physicae* and the *Exercitationes* in terms of an evolution of ideas. What has evolved, however, is the structure of the argument. The *Idea physicae* invokes all kinds of beings (*entia*), including atoms, without properly introducing or justifying them. The *Exercitationes*, by contrast, with its extensive introductory ontology, epistemology and logic carefully prepares the territory. It is as if Gorlaeus had understood, or had been told, that the novel physics he presented in the *Idea physicae* required a metaphysical basis and that he therefore decided to compose his more extensive *Exercitationes*. That the latter work is posterior in time may also be concluded from the fact that the *Exercitationes* is clearly unfinished, while the *Idea physicae* is complete.

Logically speaking, the relation between the two works is this: The *Idea physicae* stands to the *Exercitationes philosophicae* in a relation of subordination, even though it occasionally expresses Gorlaeus' ideas with greater clarity and succinctness.

For our presentation, this means that Gorlaeus' overall project is best presented by following the philosophical exposition given in the *Exercitationes*, using the *Idea* as an auxiliary text. Concerning the chronology of composition, the two works must both have been written between 1610 and early 1612, because they both contain a reference to Galileo's astronomical discoveries announced in the *Sidereus nuncius* of 1610.[3] Nevertheless, I tend to think that the *Idea physicae* was written first, the *Exercitationes* later. This assumption is not only prompted by the just-mentioned fact that the *Exercitationes philosophicae* is unfinished, while the *Idea physicae* looks complete. There are also some differences in content that suggest that Gorlaeus first worked on his thematically more restricted short treatise on physics, before starting work on his more ambitious *Exercitationes*, which prefaces the physical part with an extensive metaphysics. Such an order would also follow the logic of his biography, as we shall be able to verify in chapter 3. However, irrespective of whether this chronology is correct or not, the extreme temporal vicinity of the composition of the two works and the near total doctrinal identity between them allow us to treat them as the expression of an almost unchanged state of mind.

The method followed for our paraphrasis is the typical mixture of historical and rational reconstruction that most intellectual historians tend to use almost instinctively. The term 'historical reconstruction' refers to the historian's attempt to stay as close to the author's viewpoint as possible, by expounding what the historical author seems to have found important (rather than what we find important or interesting from our own perspective) and by employing his own terms (rather than ours). 'Rational reconstruction', by contrast, refers to the contribution that the historian makes to the historical argument in the process of organizing and expounding it. It also refers to the evaluative and critical elements that enter into his presentation. After all, we do not simply want to parrot Gorlaeus' words; we are also entitled to point out apparent difficulties or opacities in his argument and explain them either by conceptual means or through external influences such as religious, ethical or political circumstances.

Finally, wherever we do not understand what Gorlaeus means, or why he proposes what he does, we will follow a method, sketched by Quentin Skinner, that might be characterized as 'concentric': one elucidates texts by placing them in the context of other texts, first by the same author, followed by his acquaintances and friends and then by contemporary authors with which he was acquainted. Finally, one places all of these texts in the socio-cultural context of the period in which the text one wishes to understand originated. The meaning that one thereby believes one has uncovered must, however, never be identified with the intention of the author. Inevitably, the author's intention will remain disappointingly elusive. However, what can be gauged is the significance of certain ideas in a given situation, notably by the reactions they elicited.[4] Judging by this method, Gorlaeus' ideas had a larger impact than has hitherto been understood.

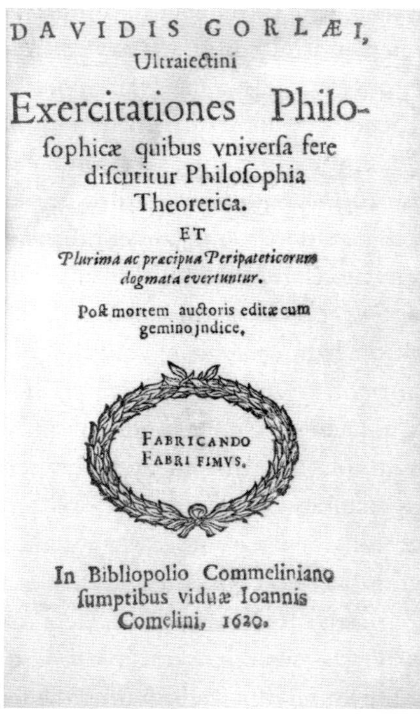
DAVIDIS GORLÆI,
Ultraiectini
Exercitationes Philoſophicæ quibus vniverſa fere diſcutitur Philoſophia Theoretica.
ET
Plurima ac præcipua Peripateticorum dogmata evertuntur.
Poſt mortem auctoris editæ cum gemino jndice,
FABRICANDO FABRI FIMVS.
In Bibliopolio Commeliniano ſumptibus viduæ Ioannis Comelini, 1620.

FIG. 3: Almost all known copies of Gorlaeus' *Exercitationes philosophicae* (1620) indicate "the widow of Jan Commelin" as the editor of the book. (Courtesy of Tresoar, Frysk Histoarysk en Letterkundich Sintrum, Leeuwarden)

2.2. A DESCRIPTION OF GORLAEUS' TWO WORKS

Let us begin with a description of the two posthumous works. The first and longer of the two is an octavo edition of 352 pages, which carries the following information on its title page (see Figures 3 and 4):

> Philosophical Exercises of David Gorlaeus of Utrecht, edited after the death of the author, in which the entire body of theoretical philosophy is discussed, and in which several essential dogmas of the Aristotelians are overturned. With a double index.

All but one out of the dozens of copies of the *Exercitationes* that I have examined carry a title page that corresponds to Figure 3, which indicates as the publisher of the book the Commelin firm and a commission by the widow of Jan Commelin, without mention of the place of publication. There exists one copy indicating a different publisher (see Figure 4): it is the copy of the British Library, which reads: "Leiden, commissioned by Jan Ganne and Harman van Westerhuyzen, 1620."[5]

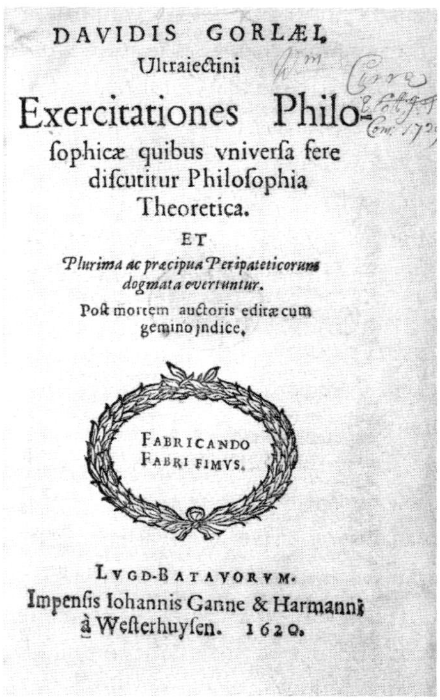
DAVIDIS GORLÆI,
Ultraiectini
Exercitationes Philo-
ſophicæ quibus vniverſa fere
diſcutitur Philoſophia
Theoretica.
ET
Plurima ac præcipua Peripateticorum
dogmata evertuntur.
Poſt mortem auctoris editæ cum
gemino jndice.
FABRICANDO
FABRI FIMVS.
LVGD-BATAVORVM.
Impenſis Iohannis Ganne & Harmanni
à Weſterhuyſen. 1620.

FIG. 4: There is only one copy of the *Exercitationes philosophicae* currently known that indicates Leiden as the place of printing and Jan Ganne and Harman van Westerhuyzen as the publishers. (Courtesy of The British Library)

The publisher, who appears on the overwhelming majority of copies as Jan or Johannes Commelin or Comelinus, Jr. (1548-1615), had transformed himself from a printer into an international publisher with offices in the Netherlands, Heidelberg and Geneva. After his death, his widow, Trijn Jansdr. Valckenier, continued to run the company, probably out of Amsterdam as some sources suggest, until her own death in 1621.[6] In the same year of the *Exercitationes*, she commissioned other publications, which carry the same indication as that found on Gorlaeus' title page ("In Bibliopolio Commeliano," etc.), including an edition of Cato's *De re rustica* and the seventh edition of Augustin Marlorat's *Expositio* of the New Testament. However, in those cases, she used the Commelin's printer's device, a naked woman in a laurel wreath, which is absent in the case of the work under examination. The emblem resembles that used until 1619 by Jodocus van Coster, but the motto inserted therein ("Fabricando fabri fimus") is found on no other publication of the period; neither by Van Coster nor anyone else.

What might the link have been between the Commelins and the set of names found on the London copy? Around 1620, Jan Commelin's sons Abraham, Isaac and Jacob tried to get a foot into the Leiden book business. Their mother, "Commelin's widow," as she is called on Gorlaeus' title page, helped them in this attempt. The *Exercitationes* were probably financed by her for the greater part, with the Leiden publishers and printers Johannes Ganne and Harman van Westerhuyzen as junior partners in this enterprise and the Commelin sons as intermediaries. That Ganne had contacts with the Commelins is known from legal acts. That Ganne and Van Westerhuyzen figure only on a single known copy of the book might in this case have to be explained by the fact their financial contribution entitled them only to a small fraction of the total number of copies printed and sold.[7]

Little is known about Ganne and Van Westerhuyzen, who worked at the lower end of the spectrum of Leiden printers; earning their living with small-scale publishing, printing and book binding. Most survey works on Dutch seventeenth-century publishers and printers simply ignore the two men, although it appears that Van Westerhuyzen was the more active between them.[8] There exist a number of editions of Dutch poems by the famous humanist and Leiden professor Daniel Heinsius (1580-1655) on which his name appears as a publisher.

Let us however return to the title page, which, as we have seen, is dry and factual: it gives the author's name and birth place, indicates that the publication is a posthumous work, and gives a descriptive title which tells the reader what he can expect to find in the book. Whether the title is Gorlaeus' own or has been added by its editor is unclear. It has the same logic, however, as the title that Sébastien Basson had chosen for his atomist treatise of natural philosophy of 1621: "Twelve books of natural philosophy against Aristotle. In which the forgotten physics of the ancients is re-established and Aristotle's errors are refuted by solid arguments."[9] We may assume that at least the main title, *Philosophical Exercises*, corresponds to Gorlaeus' intention, because the eighteen chapters that structure his book are called "exercises." This preference for "exercises" over "chapters" may in fact contain an homage to Julius Caesar Scaliger's polemical "Fifteenth Book of Exoteric Exercises about Subtlety, against Girolamo Cardano."[10]

Julius Caesar Scaliger (1484-1558), who was born as Benedetto Bordone into simple circumstances, claimed that he was a scion of the Della Scala family, which had ruled over Verona for a century and a half. Flamboyant in his life and in his work, he trained in philosophy and medicine and worked for most of his life in France. His extremely disorganized but brilliant *Exotericae exercitationes* of 1557 (of which only the fifteenth volume was published!) enjoyed a great but to some degree still ill-understood success north of the Alps, where it was reprinted frequently well into the seventeenth century. These polemical exercises pretended to be fiercely conservative in defending Aristotle against modern ideas such as Cardano's, while in truth proposing a host of new ideas.[11] As for Gorlaeus, that he admired Scaliger and particularly his *Exercitationes* is evident: Scaliger is the only recent author who is mentioned by name in both of Gorlaeus' books. Moreover, he is also invoked in Gorlaeus' entry in the *Album amicorum* of his friend Engelbert Egidius van Engelen.[12]

By contrast, the editor of the *Exercitationes* remains unnamed. Until new evidence comes to the fore, we must leave this question unanswered. It clearly cannot have been Gorlaeus' parents, as they had both passed away in the meantime. But who else had an interest in 1620 to publish the reflections of a student who had passed away eight years earlier, and to do so at that precise historical moment? Was it simply the executor of the testament of Gorlaeus' father, Carel van Gelder, who was also Gorlaeus' cousin, who decided to have the book printed for a profit?[13]

If the reconstruction of the intellectual intentions of the *Exercitationes* provided below is of any value, it is clear why this work would have been published in such a secretive fashion: to readers who had either known Gorlaeus personally or who understood the philosophical and theological thrust of his philosophy, it would have been evident that it was connected to the Remonstrant (or 'Arminian') faction of Dutch Protestantism, which had officially been condemned at the Synod of Dort the year before the *Exercitationes* was published.[14] Knowing about the Gorlaeus family's connection to the Arminians, the Dutch chemist and historian Frans Maurits Jaeger long ago proposed that the philosopher and Arminian theologian Petrus Bertius, an acquaintance of Gorlaeus' uncle Abraham, might have been the editor of the *Exercitationes*.[15] This suggestion was however based on the mistaken idea that Gorlaeus had taken his first degree at Leiden's *Statencollege*.

Finally, it must be pointed out that the process of editing, publishing and proofreading were carried out hastily and negligently. To begin with, Gorlaeus, who seems to have been a competent Latinist – as is evidenced by both his childhood poetry and his *carmen* in praise of Stellingwerff – does not give his best in this work, which is written in pedestrian and oftentimes wooden Latin. No one seems to have edited his sometimes sloppy style before publication. Nor did the printer and his proofreaders do a much better job. The thoroughly corrected copy held at the Universitätsbibliothek Basel, for example, shows how often a competent reader could have reason to take exception with the printed text.[16] The only improvement to the manuscript may have been the *index rerum* that was added to it. The most severe shortcomings of *Exercitationes* are however not only of a grammatical, syntactic or typographic order. The most obvious defect of this work is that it is incomplete. The book's eighteenth and last "exercise" is not only unfinished, but is clearly tacked on.

As for Gorlaeus' second treatise, the *Idea physicae*, it is an exceptionally rare book. Although it was printed in Utrecht, no public collection in the Netherlands owns a copy of it, and most early modern bibliographers ignored its very existence. Its title page conveys the following information (see Figure 5):

> Sketch of Physics, of David Gorlaeus of Utrecht, to which is added an Epistle by an anonymous author on the Motion of the Earth.

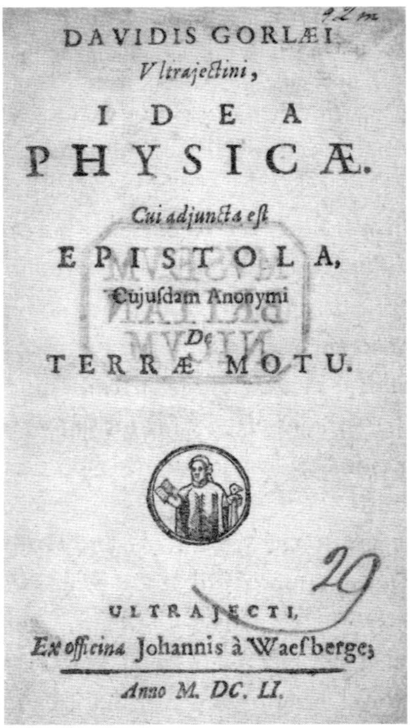

FIG. 5: Gorlaeus' *Idea physicae*, published together with Rheticus' treatise on the motion of the Earth, is an exceedingly rare book. No Dutch public library currently possesses a copy of it, although the work was printed in Utrecht. (Courtesy of The British Library)

This time, there is no mystery concerning the publisher. Johannes Janssonius van Waesberge (active between 1642 and 1659 in Utrecht and thereafter in Amsterdam) was an established printer and publisher, who worked in Utrecht from a shop that was located in front of the town hall.[17]

The book containing the *Idea physicae* is a tiny, modest duodecimo edition, of which Gorlaeus' treatise takes up only 76 pages. As the title page indicates, the book contains also a second treatise, which is called "Epistle on the Motion of the Earth" (*Epistola de terrae motu*) on the common frontispiece and "Dissertation on the Hypothesis of Copernican Astronomy" (*Dissertatio de Hypoth[esi] Astron[omiae] Copernicanae*) on the separate title page. What to Van Waesberge was an anonymous treatise was some decades ago identified by Reijer Hooykaas as a highly important early treatise by Georg Joachim Rheticus (1514-1574), Copernicus' only pupil and his most faithful early propagator. This *opusculum*, which according to Hooykaas may originally have carried the title "About the Motion of the Earth and Holy Scripture" (*De terrae motu et Scriptura Sacra*), was written between 1532 and 1541, and argued not only that Copernicus' heliocentric model was correct, but also that it agreed with the Bible, which, if interpreted correctly, included various heliocentric references.[18] Copernicus' best friend, Tiedeman Giese, Bishop of Culm, wrote in July 1543 a letter to Rheticus, in which he spoke with praise of this "little work by which you have skilfully protected the motion of the Earth from disagreement with the Holy Scriptures."[19] According to Robert Westman's recent reconstruction, there even existed plans, possibly shared by Copernicus himself, to include Rheticus' treatise as an integral part of the first edition of *De revolutionibus orbium coelestium* of 1543. "With Copernicus' death on the eve of the Council of Trent (1545-63)," Westman writes, "this brief gesture of philosophical and exegetical openness would go unheeded until second- and third-generation Copernicans independently revived Saint Augustine's principle of accommodation more than a half century later."[20]

It is highly surprising to find Rheticus' and Gorlaeus' treatises published in a single volume, all the more as the publisher did not know Rheticus' identity, whose important treatise he merely "appended" to Gorlaeus' text, as the title page indicates. Furthermore, the two authors had nothing in common. Gorlaeus was no Copernican, while conversely, Rheticus was no metaphysician, let alone an atomist.

Obviously, given the rarity of this book, Rheticus' work could not have had any greater impact than Gorlaeus' *Idea physicae*. Moreover, as Hooykaas has pointed out, "in 1651 Kepler, Galileo and a host of theologians, philosophers and astronomers, Roman-Catholics as well as Protestants, had already tackled the problem," and so Rheticus' work may have seemed outdated.[21]

The same may be said of Gorlaeus' work: by 1651, a spate of other atomistic works had been published and the northern part of Europe was in the grip of a veritable Cartesiomania.[22] Whatever novel idea Gorlaeus may have had back in 1610 would by then not only have been known from his *Exercitationes*, published three decades earlier, but would moreover have seemed pale and stale compared to Descartes' or Gassendi's metaphysics and physics. For this reason, the dramatically belated publications of Gorlaeus' and Rheticus' two treatises were in reality stillbirths.

In his short epistle to the reader, the editor explains that the two works had never been edited before and had remained enclosed within private walls, but that an eminent man had finally decided to present them to the public.[23] Who this eminent man was, and why he wished to remain anonymous, is a matter of speculation. There is reason to believe, however, that both treatises came directly from the library at Martenastate, the mansion in which Gorlaeus' parents had resided. There exists a possible family link between the Van Waesberge family and the family of the wife of Carel van Gelder, who had been the executioner of the will of David van Goorle, Sr.

Having examined the identity of the possible editors, publishers and printers of Gorlaeus' two works, we must also address an obvious question here, regarding the author himself. Given his early death, the late publication of his two works and the anonymity of their editors, why should we actually take it for granted that the theology student David Gorlaeus was really their author and that they were moreover written by the same person?

To begin with, as has been explained earlier, the two works display such a thorough doctrinal overlap and so many stylistic similarities that there is no reason to doubt that they are by the same author. But how do we know that this author was David Gorlaeus? There are five reasons to believe in the veracity of his authorship. First, as will be shown in chapter 3, there are important doctrinal overlaps between the *Exercitationes* and the *Idea physicae*, on the one hand, and the teachings of Gorlaeus' teacher and landlord Henricus de Veno of the University of Franker, on the other. Second, the final, incomplete section of the *Exercitationes* hints that the author now studied theology, as reference is made to his academic "responses given to passages in Holy Scripture."[24] Indeed, Gorlaeus had enrolled at Leiden's theological faculty a year before he passed away. Third, both works refer in a curious and chronologically helpful way to Galileo's celestial discoveries. In the *Idea physicae*, the respective passage reads as follows:

> The Milky Way is no meteorological phenomenon, but the sheen of exceedingly small stars, which because of their excessive smallness cannot be seen; this a certain mathematician from Padua testifies to have seen with the help of the newly invented telescope.[25]

Galileo's *Sidereus nuncius*, which announced the telescopic resolution of the Milky Way into a myriad of individual stars, was published in March 1610. This gives us the *terminus post quem* of the composition of Gorlaeus' two works. And yet, no one would have referred to Galileo in this manner around 1620. The passages under investigation were clearly written at a moment in which news regarding the recent telescopic discoveries reached the European centers of learning, but in which Galileo Galilei had not yet become a household name. Indeed, it is difficult to imagine that in the years following 1612 – the year in which Gorlaeus died – anyone would have referred to Galileo as "a certain mathematician from Padua," not only because everyone would by now have recognized his name, but also because he had meanwhile become court mathematician and philosopher to Archduke Cosimo II de' Medici. The fourth reason for believing in Gorlaeus' authorship of the two treatises is that the unfinished state of the *Exercitationes*, with its provisional tacked-on conclusion, fits the concept of a posthumous publication better than that of a pseudonymous work. And finally, there is that 1641 testimony, partially cited earlier, by Gijsbert Voetius, who seems to have known the author and the circumstances of the genesis of his works intimately. Taken together, these five reasons point exclusively to Gorlaeus and no one else.

Having moved these questions out of the way, let us delve into the two books themselves and try to capture the essence of the ideas they contain. For the reasons mentioned above, our paraphrase will follow the structure of the *Exercitationes* and cite the *Idea* wherever it either expresses the same thought in a more poignant way or adds something to the larger treatise.

2.3. ONTOLOGY AS "FIRST OR UNIVERSAL PHILOSOPHY"

Like all self-respecting authors of early modern philosophical textbooks, Gorlaeus first defines his terms. In keeping with the title of his *Exercitationes philosophicae*, he opens with the question, "What is philosophy?"[26] All philosophy courses at the time began with this question, and they usually pondered a number of possible answers: ancient authors, Aristotle first among them, had sometimes identified philosophy with wisdom (*sapientia*) and sometimes with the quest for first, metaphysical knowledge; but usually, they settled for a general description, as did the Wittenberg natural philosopher, physician and atomist-to-be Daniel Sennert in 1600:

> Philosophy is an acquired power of the intellect (*habitus intellectus*), made up of wisdom and prudence, which contemplates everything and governs human actions in a congruent manner, enabling man to obtain the highest good.[27]

But none of that is found in the *Exercitationes*, which offers a definition that is shockingly short and differs from the standard answer. According to Gorlaeus, philosophy is "the naked knowledge of beings."[28] His book thus begins with a veritable bang: Gorlaeus identifies philosophy with ontology, the metaphysical discipline concerning "being as such."

Abstract though such naked knowledge may seem, so Gorlaeus continues, it is of the highest importance to us as it leads us to a better understanding of God and provides an antidote against the misconceptions of our soul. Whereas theology teaches us how to reach paradise, "philosophy is the doctrine of how to perfect the human soul in this life."[29] The human soul is in fact deficient in lots of ways: certain objects it cannot grasp because of their perfection (such as God), their imperfection (such as time, space and action), their greatness (notably the world) or their smallness (notably atoms). Philosophy, if reformed properly, will help us find our way out of our mental quagmire. In this operation of philosophical assistance, metaphysics plays the key role because, Gorlaeus explains, one of the main scourges of mankind is that our intellect labors under the problem of referentiality: our notions (*notiones*) do not correspond to things (*res*) as they really are. Fortunately enough, however, philosophy can help greatly in improving this situation, notably by teaching us how to discern between existing and non-existing beings and their modes.[30] In fact, philosophy can be compared to a kind of doctor's kit, because for each human deficiency there exists a corresponding branch of philosophy that may cure it. Ethics, for example, helps us to mend the imperfections of our will, whereas the illnesses of the intellect are tackled, if practical, by mathematics, and if theoretical, by a type of "logic" that is in truth a "science of being" (*entis scientia*).[31]

As we shall see below, Gorlaeus' teacher Henricus de Veno regarded man's limited intellectual capacities as a consequence of original sin.[32] Gorlaeus does not explain our mental inadequacy in theological terms, but merely points to the inadequacy of our senses with respect to reality and the mistaken terminology that result from this mismatch. He does, however, like other authors in his day and age, subscribe to the notion that it was in man's power to improve his epistemic and moral stature through philosophy and to obtain greater perfection. This idea is for example very prominent in the work of Francis Bacon, most notably in his treatises *On the Proficience and Advancement of Learning, Divine and Human* (1605) and the *Instauratio magna* (1620). As Stephen McKnight has stressed, Bacon's program "is not a secular, scientific advance through which humanity gains dominion over nature and mastery of its own destiny. Bacon's instauration is a program for rehabilitating humanity."[33] This attempt at self-rehabilitation and, as Bacon called it, a self-governed return to "the first state of creation," smacked, as Passmore has pointed out, of Pelagianism: the heresy that attributed to humans the possibility of freely choosing between good and evil.[34] Although in Gorlaeus we encounter no comparable promise of a scientific return to pristine knowledge, his notion that philosophy can "cure" the shortcomings of our senses, perfect our souls and improve our knowledge of God takes us in the same direction. We shall see later why the theological implications of this position are important. In a nutshell, for an opponent of Gorlaeus' philosophy like Gijsbert Voetius, it was wrong to place understanding before faith or to view it as a precondition or even a helpful instrument: "Human reason is not prior, better known, or more certain than faith; hence reason cannot be its principle."[35]

As for Gorlaeus, this "science of being," which provides such salubrious medicine to the soul, encompasses in turn a number of sub-disciplines. The first of them, called *theosophia*, deals with the divine being and "the nature and attributes of God."[36] However, the premises of this discipline must not, so Gorlaeus warns us, be Aristotle's, but the Bible's. We shall see in chapter 3 that by collocating the treatment of God's nature and attributes within the discipline of philosophy, Gorlaeus took a highly controversial step, and one that would at the time be understood as a sign of his sympathies for the Arminian cause and in particular for the treatise *On the Nature and Attributes of God* of the embattled theologian Conrad Vorstius.[37]

The two other sub-disciplines of this "science of being" take care of created things: *angelographia* deals with supernatural beings, and notably angels, as the name suggests; and *physica* deals with natural beings.[38] This brief and underdeveloped division of philosophy may echo views formulated in Bartholomaeus Keckermann's *Scientiae metaphysicae compendiosum systema* of 1609 and Otto Casmann's *Angelographia* of 1605.

Given Gorlaeus' ontological approach to the subject matter of philosophy, it should not surprise us that his definition of *physica*, too, differs considerably from the standard Aristotelian definitions. *Physica* – a term that was at the time used interchangeably with *philosophia naturalis* ('natural philosophy') – was in textbooks from that period routinely "defined as a speculative science which studies the world of changing material things – celestial and terrestrial, animate and inanimate – culminating in the study of man."[39]

While the standard definition thus insisted that natural philosophy (or physics) studied nature inasmuch as it was subject to change, Gorlaeus' *physica* dealt with unchanging beings. "For nature is something absolute" (*absolutum quid*), our author explains; it is a rigid composition of unchanging beings.[40] Once again, it is fair to say that his is certainly quite an unusual standpoint. Aristotle, who supposed the word *physica* to be cognate to the verb '*fuein*' ('to grow'), had understood physics to be a discipline dealing with change and development. Indeed, the entire Aristotelian approach to natural objects had throughout the centuries consisted in considering all natural objects as subject to processes of mixing and unmixing, growth and decay, as well as of unfolding from a potential towards an actual state.

Gorlaeus was fully aware that his approach and his terminology were at odds with those of his contemporaries. He even admitted to his difficulty of finding a suitable name for his new, general science of being. In the end, and "for the time being," he chose to call this discipline the "first or universal philosophy."[41] He hastily added that we must not confuse this *prima philosophia* with Aristotle's metaphysics. The term 'meta-physics' refers precisely to those Aristotelian books that were to be placed after ('metà') those on physics. By contrast, Gorlaeus insists that his own 'first philosophy' precedes physics in the order of the disciplines. As far as he is concerned, traditional metaphysics could simply be expunged from the scheme of philosophical disciplines, as it referred to nothing more than to a bagful of disparate topics.[42]

Having defined the parameters of his universal philosophy, Gorlaeus begins to lay down its principles, and he does so by introducing the following two distinctions. Following a traditional division, he first distinguishes *entia realia* from *entia rationalia*, where "real beings" mean those whose being (*esse*) is rooted in their independent existence (*existentia*), while "rational beings" depend for their *esse* entirely on the human intellect. More innovative is, by contrast, his separation between the *ens per se* ("being by itself") and the *ens per accidens* ("being by accident"), also called *ens per aggregationem* ("being by aggregation"). The *ens per se* is an ultimate, self-supporting unit, while the *ens per accidens* or *per aggregationem* refers to an entity that only appears to be an independent unit while it is in fact, as its name indicates, a mere aggregate of several *entia*.

But here, we must take note of the conspicuous list of cases Gorlaeus offers us of such "accidental" or "aggregate beings": (i) a heap (*acervus*) of stones, where the contributing beings "touch each other only confusedly"; (ii) the world (*mundus*), whose beings touch each other "in some order"; (iii) water drops merging in a river, or a puddle, i.e., beings of the same quality joining to form a larger entity. Finally, and most controversially, (iv) there exist cases

> where one [entity] is intimately in the other, penetrates it, and acts through it, as the soul does in the body. Although various beings are joined here, they will yet together always form an entity by aggregation. Nor does their union change the essence of the things in such a way that from two things, a numerically single thing might come about.[43]

This is a remarkable list indeed. We must recall that Aristotle, and following him all of medieval and early modern school philosophy, had insisted on the distinction between a true mixture (*mixtio*) and a mere 'heap' (*acervus*): by 'mixture' was meant a genuine fusion of heterogeneous parts into a new homogeneous unity, which was defined by a single form (the so-called *forma mixti*); whereas a 'heap' designated a mere aggregate of spatially juxtaposed, but unmixed, parts.[44] In classifying all compounds as aggregates and by insisting on the continuing independence of the contributing beings, Gorlaeus sides with an essentially atomist conception of nature. Only an atomist will view a body of water as an *ens per accidens* made up of individual water units, rather than as an infinitely divisible continuum.

The most remarkable aspect about Gorlaeus' list of *entia per accidens* is however his definition of man as a composite being. As we shall see in chapter 4, it is a definition with a remarkable earlier and later history. Gorlaeus took it from Nicolaus Taurellus' *Triumphus philosophiae* of 1593, while Henricus Regius, in turn, was to lift it out of Gorlaeus several decades later, letting a student defend it in a disputation in 1641. This triggered Gijsbert Voetius' violent anti-Cartesian reaction, which led to the prolonged Crisis of Utrecht and to the official condemnation of Cartesianism.[45] It is a tenet that Pasnau has called a "shockingly explicit version of Platonic dualism."[46] Although Gorlaeus nowhere cites Plato, his view entails, in his own words, that

> in a human being, there is a soul and also a body, and these two are united in such a way that the body is made the soul's residence, vehicle, and instrument through which the soul exercises its operations. But these two are not made into one being, called a human being. Instead, each retains its complete and perfect essence, by which it is what it is. Still, the human being is not the same as the soul, nor the same as the body; rather, it is the same as the soul and the body taken together and aggregated. If, however, the human being is to be considered not as a being by aggregation, but as one thing *per se*, then it will be the same as the soul existing in the body.[47]

The reasons that may have persuaded Gorlaeus to accept Taurellus' view will be discussed below. But one theological argument is offered in the *Exercitationes*. It has to do with death and resurrection: from Sacred Scripture, I learn that

> I wish to dissolve and be with Christ. But if the composite [of body and soul] is one being, and man is a composite, then it is necessary that in death, man's essence perishes, because that being which is a composite disappears. How, then, can that which is not a being be with Christ? And how could death be called a dissolution, if it is instead a corruption of the substance? But should one call disappearance, what is merely dissolved? Since this is clearly not a coherent view, we bid farewell to this Peripatetic nonsense and follow the sacred truth.[48]

We have encountered this very Platonic idea of man as a merely temporal composite of soul and body already on Gorlaeus' tomb! "This very illustrious mind and heavenly spirit, / Freed from its fetters and the weight of the body, / Sought the Havens, whence had come its seed."[49] One wonders whether the tombstone and Gorlaeus' argument here are in any way related; was the author of the epitaph familiar with Gorlaeus' philosophical views? Did Gorlaeus write his own epitaph?

But we must return to the argument. It is obvious that Gorlaeus adamantly defends a view according to which nothing ever changes. Souls do not change, atoms do not change; and when they enter into a temporary collaboration, forming human beings for the duration of their lifetime, they certainly do not merge into something different but remain distinct units. According to Pasnau, this "strict permanence thesis [...] seems to have been given its first sustained post-scholastic defense by David Gorlaeus," who wished to replace Aristotle's hylemorphism "with an atomism that insists on strict permanence."[50] Indeed, unless God creates something out of nothing, there is no way in which something new can come about: "I deny," Gorlaeus writes, "that any body has been changed into another, or that it can be changed."[51]

The straight link between the imperishable and immutable *ens per se* and Gorlaeus' atomist ontology becomes even more evident when we look at the general description the *Exercitationes* offers of the ultimate unit of being (*ens*). Each *ens*, its author decrees, must exist actually (*in actu*) – unlike Aristotle's substances, which can also exist potentially (*in potentia*). Indeed, Gorlaeus decries all scholastic talk of potency as muddle-headed.[52] In this, he is the radical heir of a long tradition that ultimately goes back to William of Ockham (c. 1288-c. 1348), for whom an *ens* was a concrete singular thing (*ens singularis*). Two centuries later, Lorenzo Valla (1407-1457) insisted that "being" (*ens*) was the same as "what is"; "what is" was the same as "that which is"; and "that which is" what the same as "this thing."[53] Again almost two centuries later, we now find Gorlaeus insisting that all *entia* possess unchangeable essential properties, and that existence is one of them. In fact, "essence" and "existence" coincide.

Although the label 'nominalism' had a different meaning for each generation ever since it was first used in the fifteenth century, Gorlaeus clearly thought of himself as a nominalist. He self-confidently battles for this view with what he took to be nominalist weaponry, exclaiming (according to F.A.H. Peeters' scrupulous count) a full seventeen times in the *Exercitationes* and five times in the *Idea physicae* that "beings must not be multiplied beyond necessity."[54] He valiantly, and sometimes recklessly, wields Ockham's razor as an "Achillean defender of our doctrines."[55] As far as the issue of universals is concerned, his self-description as a nominalist is certainly correct, according to the standards of his time: "All that exists *in re* is one by number and singular," he insists. Universals, by contrast, are mental abstractions, which establish their reference through what would nowadays be called 'indexicals', that is to say, by pointing to a singular "this, here, now."[56] In the material realm, these units are physical atoms in the traditional, Democritean sense.[57]

In his *De principiis et originibus*, which was written in the 1610s but published in 1653, Francis Bacon wrote:

> But a principle is not an entity, and a mortal entity is not a principle. A clearly irresistible necessity therefore drives men's thoughts (if they want to be consistent) to the atom, which is a true being (*ens*), having matter, form, dimension, place, resistance (*antitypia*), appetite, motion and emanation. Likewise, amid the destruction of all natural bodies, it remains constant and eternal. For since the corruptions of the greater bodies are so many and various, it is absolutely necessary that that which remains as an unchanging center should be something either potential or extremely small.[58]

Gorlaeus' *ens* has a number of properties in common with Bacon's, but there are also various important differences. First, from the way in which it is introduced in the *Exercitationes*, Gorlaeus' atomism is in the first instance metaphysical. Moreover, his category of *ens per se* not only includes material atoms, but also God and angels. Bacon's atomic *ens* designates that unshaken material unit that survives unharmed whatever vicissitudes the larger physical bodies are subjected to, but which also possesses active, indeed vital properties. Gorlaeus, by contrast, presents us with an entity that can be divine, spiritual or material. Obviously, the inclusion of non-material beings cannot but have profound implications for the category of *ens per se*. For whenever our author goes about defining its qualities, he invokes what might legitimately be called a 'God criterion': any attribute of the *ens*, even when this *ens* is a material atom, must also hold true for God.[59] The application of this 'God criterion' explains the choice of the following essential properties of all *entia*: "unity, truth, goodness, existence, locality, durability."[60] But obviously, the 'God criterion' also works the other way around: what is true of the atom must also hold true for God. The resulting physicalization of God is by no means an innocent or unwanted consequence of Gorlaeus' metaphysics. We shall see below, in chapter 3, that it is in accordance with the theologian Vorstius' controversial ontological treatment of God and his attributes.[61]

As for 'unity', Gorlaeus takes this property to include 'quantity', too, which in his eyes is not an accident but "is the same as substance," because both *quantitas* and *ens* are rooted in numerical unity. "Just as unity is not distinct from what is one, so the quantity of the atom is not distinct from the atom itself."[62] When three atoms join up, they therefore form a ternary number, not a single continuum. Just like Giordano Bruno a few years before him, Gorlaeus thus understands by the "quantity of an object" nothing over and above the number of its atomic components.[63]

Besides these essential qualities, beings also possess contingent *modi*, which describe their disposition (*habitudo*) within the parameters of time and space (rest, speed or duration) and also their relative spatial position as, for example, the location of a given atom with respect to the others.[64] Helen Hattab has recently individuated in Gorlaeus' replacement of the scholastic attributes by *modi* a direct model for Descartes' similar enterprise.[65] She concludes:

> Unlike prior anti-Aristotelian philosophies such as those of Bruno, Telesio, and Basson, Gorlaeus' metaphysics involves the complete rejection of the Scholastic Aristotelian substance/accident ontology in favor of a substance/mode ontology. [...] [H]is account of modes and his identification of substance with extension/quantity logically commit him to the view that all the properties of body are modes of extension well before Descartes develops his metaphysics.[66]

While the *modi* thus defined seem useful for the description of a complete atomistic system, Gorlaeus also introduces a further, somewhat unexpected and in fact rather awkward category, namely "real accidents" (*accidentia realia*), a type of *ens* that has "less perfect being" because it leads a merely semi-independent life.[67] 'Real accidents' must always inhere in a subject (which is why they are 'accidents'), but they can also migrate from one subject to another (which is why they are both 'real' and also *entia*). They are furthermore endowed with limited causal powers, for they can produce other accidents. In fact, Gorlaeus explains a great number of physical phenomena – color, rarefaction, lightness, etc. – in terms of the "accident of heat," which multiplies itself across contiguous elementary atoms.[68]

Although one may encounter 'real accidents' in various early seventeenth-century philosophers, it is hard to deny that this theory of 'real accidents' disturbs the coherence of Gorlaeus' atomistic explanations.[69] One intuits that our author simply wants to eliminate Aristotelian potencies or powers, and to redefine qualities as the essences of autonomous beings: "Just as we have said that quantity does not differ from the quantifiable body, so we also say that no potencies, indeed, no properties are distinct from the essence of a thing. They differ solely by our reason and manner of conceiving them."[70] This view has been described as a "nominal conception of power," in the Lockean sense of the word, meaning that "the substance itself is capable of whatever operations are ascribed to the powers."[71] But whereas most later seventeenth-century corpuscular thinkers, and certainly John Locke, would follow in the footsteps of Plato's *Timaeus* and explain sensorily perceived, so-called 'secondary' qualities in terms of the geometrical or 'primary' shapes of corpuscles, Gorlaeus' doctrine of migrating 'real accidents' bars this type of geometrical reductionism and the mechanical understanding of sense perception that usually accompanied it. Surprisingly, in fact, our philosopher displays the utmost indifference with respect to the explanatory possibilities offered by postulating specific atomic shapes, preferring to "leave this question unresolved."[72] For him, the two most important qualities "inhere" in the atoms, of which they constitute the essence, while the others are the result of the structure formed when several atoms gather.

These two essential qualities are humidity (for water) and dryness (for earth). They are not due "to the congregation of atoms, but exist in the atoms themselves: for if the singular atoms were not dry, the entire body could not be dry."[73] Hot and cold, Aristotle's other two primary qualities, are by contrast 'real accidents'. They migrate from one subject to another and are therefore the essential properties of none. All remaining qualities are mere *modi*, to be explained as the product of atomic aggregates.

In contrast to the migrating 'real accidents', which are hard to integrate into any kind of atomist program, however defined, these *modi* take us in the direction of molecular notions developed fully by other seventeenth-century authors. "Rarity, density, lightness, roughness are similar *modi*, not real qualities," Gorlaeus writes, "for they are nothing else but the positions of parts."[74] Interestingly, when he allocates different properties to these two different levels of concretion – atoms *per se* and conglomerates *per accidens*, as it were – Gorlaeus invokes chemical experiments, of which he writes that they manage to change only certain qualities, but not others.[75]

2.4. GORLAEUS' PHYSICAL ATOMISM

"The [exercises] regarding general being (*ens in genere*) and its accidents, with which we have so far been dealing, have been fairly exhausting and difficult," we read on page 221.[76] Only in the twelfth exercise, that is, only two-thirds through his treatise, does Gorlaeus leave the explanation of his ontological 'first philosophy' so as to enter into the field of physics proper. Understandably, historians of science have focused on the last third of the book, but our summary should have made it evident that Gorlaeus' physical atomism presents itself as the logical by-product of his general atomist ontology.

The physical part of his treatise begins with a refutation of the two constitutive principles of Aristotelian hylemorphism, namely matter and form, and with their replacement by unchangeable corporeal atoms.[77] The existence of these atoms, Gorlaeus declares, cannot only be logically derived from metaphysical principles, as has been done in the first two hundred pages of his work. Atoms can also on rare occasions be perceived by our senses. For example, "who would not believe that those minute parts of water vapor are indivisible?"[78] As for Aristotle's counter-proposal to atomism, which relied on substantial forms possessing the power to turn different materials into a single homogeneous substance, Gorlaeus protests that it contradicts all principles of good philosophy. First of all, "which reason forces us to multiply beings unnecessarily?"[79] Once again, it is Ockham's razor that is employed to shave away the Peripatetic vocabulary. At the same time, however, Gorlaeus is fully aware of Aristotle's main argument against the possibility of physical atoms, which is that they put an arbitrary limit to the divisibility of an extended body. He responds to it rather elegantly by distinguishing between 'quantity', on the one hand, and 'divisibility', on the other, and thus between physical atoms and geometrical lines.[80] Throughout the seventeenth century, atomists and anti-atomists quarreled over the legitimacy of separating physical from mathematical divisibility. Take, for example, the alchemist Andreas Libavius (1555-1616), who in his *Alchemia triumphans* of 1607 claimed, albeit in an intellectually reckless manner, that Democritus had been a chemist and that his atoms referred to the principles found in the chemical resolution of bodies. To his detractors who retorted that all extended bodies had to be indefinitely divisible, Libavius replied that they failed to grasp the difference between mathematical extension and the atomic make-up of natural substances.[81] In the eyes of seventeenth-century natural philosophers, the problem remained, however, unresolved. While the atomist Pierre Gassendi, a number of decades after Libavius, accepted this distinction, his countryman René Descartes wouldn't, positing the indefinite divisibility of all matter and therefore also of the corpuscles that he invoked to explain physical phenomena.[82]

It has been mentioned above that according to Gorlaeus' natural philosophy, there exist only two essential qualities, namely, dry and humid. Aristotle's system, by contrast, features four such qualities, in addition to dry and humid, there are also hot and cold, and their binary combination explain the existence of precisely four elements. Fire is defined by the qualities hot and dry, earth by dry and cold, water by wet and cold and air by wet and hot. These four elements are furthermore defined by their motion: whereas earth is heavy and tends downwards and fire is light and rises, water and air lie in between. There is also that fifth element, ether, which has no upward or downward drive, but moves circularly: it is the element that defines the superlunary sphere of eternally regular circular motion.

As we shall see presently, Gorlaeus does not only reject this ether, and with it the venerable distinction between sublunary and supralunary regions, but his elimination of the two primary qualities of hot and cold entail a reduction of the sublunary elements from four to two. In fact, he only acknowledges earth and water as elements, devoting a considerable number of pages to eliminating the other candidates. His main reason for maintaining a two-element doctrine is this:

> Elements are said to be those things out of which the mixed bodies are composed and into which they are again reduced. [...] But we experience that the mixed bodies, which perish, are resolved into earth and water. [...] For there is no mixed body that is resolved into fire or air.[83]

More specifically, fire cannot mix with water, nor do we eat and digest it; for these reasons, it cannot be a component of our bodies.[84] In reality, fire is but an accident: it can be caused, for example, by the friction of closely packed atoms. That it is nothing but heat can be proven by the fact that when the sun shines, the air gets warm, and when we bundle sun rays through a burning glass, fire will manifest itself "to sight and touch."[85] Nor is air an element.[86] In contrast to fire, however, Gorlaeus does not place air in the category of accidents but accepts it as a real substance – which, as we have seen before, fills the entire cosmos. It is of course more than doubtful whether his ontology allows for any such type of simple non-elementary substances. The reason for its unique and very peculiar status is that, while being an independent sort of substance, it cannot mix with either water or earth, because "air cannot depose its secondary qualities and assume others."[87] Such a capacity, as we have seen, constitutes for Gorlaeus the precondition for mixing. However, air is found in the pores of all mixtures, where it retains its nature intact. As a matter of fact, our author needs air not only as a universal filler of void spaces (because he allows for no vacuum), but also as the ubiquitous carrier of the 'real accident' of heat. By nature neither hot, cold, wet or dry, air passes celestial heat down to the two elements of water and earth, thereby triggering off mixtures, generations, and corruptions.

So, what exactly happens according to this model when the only two real elements, namely atoms of earth and water, mix?

> We believe that each and every part has its own essence before entering into any composition, and preserves it in it, and that [in mixture] no numerically single entity is produced, or that one *ens* is made out of them, but that they unite and mix so that one continuum is produced, which is one aggregate being (*ens per aggregationem*), but not by essence.[88]

This is not a particularly clear standpoint. On the one hand, Gorlaeus is here found arguing against Aristotle's definition of mixture as the production of a new homogenous substance. On the other hand, he also seems to betray the basic intuition of atomism by claiming that "one continuum is produced." How exactly should we imagine a mixed body that is at the same time an *ens per aggregationem* (since the contributing atoms retain their separate identities) but also forms a continuum? His answer is as follows:

> For the minimal bodies are the atoms, which are mixed in various ways. These must touch each other mutually. For if they did not touch each other, how would *one* body arise [from them]?[89]

Mixing, in other words, is both mutual contact *and* union. When atoms touch each other, they exchange their respective qualities with one another, bringing about a shared set of properties. This sharing, in turn, guarantees the appearance of the mixt as a 'continuum'.[90] In proposing this view, Gorlaeus explicitly followed in the footsteps of the well-known sixteenth-century Italian physician and natural philosopher Julius Caesar Scaliger (1484-1558). Scaliger, who was loved as much by Protestant scholastics as by their detractors, was incidentally the father of the equally famous Joseph Justus Scaliger, the humanist and professor at Leiden University (1540-1609), who died only three years before Gorlaeus. Scaliger *père*'s very popular and frequently reprinted *Exercitationes exotericae* of 1557 purportedly defended Aristotle's physics against the pernicious innovations of his colleague and rival Girolamo Cardano; in truth, however, it introduced a whole range of conceptual innovations. Among the most successful innovations was his novel definition of mixture.

Aristotle had provided a definition of mixture that had become canonical. In Latin, it reads: *Mixtio est miscibilium alteratorum unio*.[91] Literally, this means: "Mixture is the union of mixable bodies that have been altered." According to Ingemar Düring, such a literal translation does not do justice to what Aristotle wished to convey. In fact, these "five words require 19 words to render them: 'Chemical combination is a unity of bodies, capable of such combination, whose constituent parts have undergone a thorough transformation.'"[92] The result of this chemical 'transformation', at any rate, is a new substance characterized by its specific form, the so-called 'form of the mixt' (*forma mixti*).

In his very influential reformulation of that definition – a reformulation that incidentally possesses an as yet unstudied prehistory in medieval medicine – Julius Caesar Scaliger stated that mixture was "the motion of minimal bodies towards mutual contact so that a union comes about."[93] Many early-seventeenth-century atomists, including Sébastien Basson, Daniel Sennert and Joachim Jungius, eagerly seized this definition and gave it an atomist twist. This is exactly what Gorlaeus does, too – but at an earlier date – when he first quotes Scaliger's definition and then adds the following clarification: "By minimal bodies I mean indivisible atoms."[94]

This reliance on Scaliger, however, comes at a certain cost. His notorious doctrinal inconsistencies notwithstanding, Scaliger was no atomist, but, if anything, a minimist. The technical niceties of this distinction need not detain us here; suffice it to say that those early modern scholastics who developed Aristotle's sketchy and controversial allusions to 'natural minima' into an explanatory tool did normally not think of these minima as atoms; that is, as independent, self-subsisting beings. Rather, they understood minima as the lowest limit of a quantity of matter capable of maintaining a certain substantial form. Put differently, they regarded both natural minima and maxima as limits, not as things. For this reason, they posited a natural minimum for every substance.[95] Gorlaeus, by contrast, like all genuine atomists, could only permit the existence of as many types of atoms as there were basic substances and elements; of which, as we have seen, he acknowledged three, namely the two elements of earth and water, and the non-elementary substance of air. For this reason, his attempt to build on Scaliger's doctrine of mixture and of minima took the risk of ushering in a confusion between the different types of logical reasoning behind atomism and minimism. And in fact, Gorlaeus occasionally gets caught in this conceptual trap, for example when he declares that there are minima of sand just as there are minima of water.[96] If you are a minimist, this is of course correct; but if you subscribe to Gorlaeus' doctrines, it is not, as you would have to define sand particles not as *entia per se*, but as merely accidental mixtures of earth and water atoms.

The problem of the explanation of mixture and the new properties that emerge in the process of mixing takes us straight to a problem concerning which both Aristotelian natural philosophy and atomism seemed to be at a loss. The issue is the same, irrespective of whether one postulates four elements, as the Peripatetics did, or a small number of corpuscular types, as Democritean atomists did, and it is the following: if one assumes that mixed bodies are the result of a combination of no more than four elements, how is one to explain that these display so many different and such radically new properties? How can earth, water, air and fire mix once into drinkable milk, another time into poisonous arsenic, life-giving blood or unbreakable diamonds? Aristotle had proposed that the mixture of elements implied the blending of the various elementary qualities and the concomitant emergence of what he called a common 'temperament'. However, his explanation seemed insufficient to clarify, for example, how it was possible that blending non-toxic elements could possibly end up giving rise to a toxic 'temperament'. The superaddition of a substantial form (*forma mixti*), which characterized the new homogeneous mixture, was therefore required. But where did this new form come from? Was it the automatic result of the *temperamentum*, or was it in some other way 'educed from the matter', as this process was sometimes called? Or was it instead in some way dependent on external influences such as the heat of the sun or stellar rays? This question, which Aristotle himself left in the middle, remained at the center of a drawn-out, yet unresolved, controversy; so much so that in the course of the seventeenth century, the very concept of a *forma mixti* could become a matter of ridicule and a cherished proof of the vacuity of scholastic terminology.[97]

But to laugh at the *forma mixti* was easier than to replace it. Seventeenth-century atomists, who in this were partly following the lead of such ancient atomists as Lucretius, felt forced to invent hooks and spirals, minuscule chains and wedges as well as magnetic poles in equally futile attempts to explain such emergent properties as liquidity, sweetness, redness, toxicity and so forth by means of adventurous combinations of geometrical shapes.

As for Gorlaeus, he attempted to address this complex problem with the conceptual tools that he thought were at his disposal. As mentioned before, he explained many of the emergent properties of mixtures in terms of the *modi*, that is to say, the reciprocal position of atoms. But he apparently sensed that he could not reduce all qualities to such merely spatial arrangements. We recall that spatial contiguity of parts was for him only a precondition of mixing. When atoms touch, so he thought, they exchange their qualities and bring about a certain qualitative union. We recall from above that his particular explanation involves 'real accidents', these travelling qualities of hot and cold, which are said to interact with each other in such a way that one 'middle quality' or *temperamentum* is produced among the intermingling atoms.[98] Whenever outside influences disturb this 'temperament', the body will once more resolve into its original components.[99] In sum, then,

> in mixture, heat and cold function as the efficient cause, while humidity and dryness are the material cause. And thus arises the temperament, which is nothing else than the due proportion between their four qualities.[100]

This doctrine recycles the traditional idea of the common *temperamentum* arising in homogeneous mixts, only that this 'temperament' is now explained by means of spatially contiguous atoms sharing qualities and real accidents rather than by means of elements dissolving into a new continuous mixt. Several commentators have viewed this doctrine as a sign of half-heartedness and of a pusillanimous quivering half-way between scholasticism and a fully mechanical conception of nature.[101]

This criticism is, in my view, mistaken. Had he wished to do so, Gorlaeus could have easily subscribed to the geometrical type of atomism that Aristotle or Diogenes Laërtius sketched in their descriptions of the philosophy of Leucippus and Democritus, which Epicurus expounded in his extant fragments, and which Lucretius had exhaustively illustrated in his *De rerum natura*. If he hesitated to do so, this should not be explained merely by his membership in an allegedly transitional generation of thinkers who had not yet collected enough mental strength for a full rupture with 'the Philosopher'. Rather, one should keep in mind that not one single seventeenth-century 'atomist' was fully persuaded by the atomistic model provided by the ancient sources. The first atomist to make this clear was Giordano Bruno, who on the one hand imitated Lucretius' verse and made engravings to illustrate the stacking of Democritean atoms, but who on the other hand explicitly stated that "atoms and the void are not enough" to explain natural phenomena and endowed his atoms with vital forces.[102] The same holds true of Francis Bacon, who, despite his admiration for Democritus' philosophy, felt that the properties and behavior of bodies could not be explained through their shape, motions, collisions and entanglements alone.[103] By the same token, even when turning his back on Aristotle, the German physician, philosopher and chemist Daniel Sennert did not espouse a materialistic version of atomism, but decided to maintain the old substantial forms which he now located in his atoms.[104] With respect to his doctrine of mixture, I think we have reasons to comprehend rather than to deplore Gorlaeus' reluctance to embrace the naked geometrical materialism of ancient atomism. Like most seventeenth-century atomists, Gorlaeus too felt that the generation of new qualities in chemical mixture had to involve more than a merely spatial gathering of atoms, and that the atoms contributing to a mixture must at least be partially transformed and homogenized so as to generate the "qualities that bind the parts together in unity."[105]

Let us return now to the overall scheme of Gorlaeus' theory of the elements. We have just seen that in what used to be Aristotle's sublunary world of 'generation and corruption', he abolishes two of the four traditional elements and redefines mixture in atomistic terms. Even more radical, however, are the consequences of his decision to abolish also the fifth element, ether. In what contemporaries perceived as an important doctrinal novelty, he combines Holy Scripture with evidence provided by the sciences to rule out the existence of the ether and to insist that the whole cosmos is filled with the same matter. In doing so, he denies the central Aristotelian distinction between two physical realms, namely a sublunary world defined by the four elements, rectilinear upward and downward motion as well as by constant coming-about and perishing, and a supralunary realm defined by the element ether, by circular motions and by the constancy and incorruptibility of the planets and stars, which were attached to crystalline spheres. The elimination of this distinction constituted one of the most important ingredients of, and indeed preconditions for, the advent of the seventeenth-century's new sciences. In this respect, Gorlaeus shows himself to be aware of the requirements that physics had to fulfill in the age of Nicholas Copernicus, Tycho Brahe, Johannes Kepler and Galileo Galilei. After all, if the location of the Earth was no longer conceived to coincide with the center of the universe, but as lying on a planetary orbit around the sun, the traditional layering of elements up to the moon no longer made any sense. In fact, one of the most glorious chapters of seventeenth-century science between Kepler and Newton contains the story of the cosmologically driven overhaul of physics and astronomy.

Gorlaeus' rejection of the difference between sublunary and supralunary physics is an integral part of this glorious chapter in the history of science. In fact, that he put it down on paper as early as c. 1611 is remarkable, since most Dutch university courses continued to teach Aristotelian cosmology for some decades to come.[106] It is when dealing with this issue, more than in any other part of his book, that one notices Gorlaeus' acquaintance with contemporary research in optics and astronomy. For example, in order to prove that the heavens are not filled with ether, but with air, he invokes "the shared conviction (*dogma*) of opticians that different diaphanous media generate refraction." But since such refraction and the concomitant displacement of sun, moon and stars is only demonstrated "in the lower region of air, because of vapors and exhalations," and not elsewhere, it follows that there exists no other medium in the heavens than air.[107] Equally interesting is Gorlaeus' astronomical argument against the existence of celestial spheres, to which the planets were traditionally assumed to be affixed: "Besides, the observations of astronomers (*mathematici*) show that comets appear in the heavens themselves."[108] This remark contains an obvious reference to Tycho Brahe's measurements of the Great Comet of 1577, which proved that comets were not meteorological phenomena, occurring in the region between the Earth and the Moon, as standard theory had it, but that they traversed the orbits of the planets in the superlunary region.

But if the planets and stars are not attached to any crystalline sphere, how then do they move? Gorlaeus proposes a hydrological solution: "The heavens are a continuous body, which does not move, but is quiet. The stars themselves move in it freely, in the same way in which fish swim in water or rather how clouds are carried about in the air."[109] Although the theory of the fluidity of the heavens is found as far back as in Ptolemy's *Hypotheses of the Planets*, it had become a minority view in the Middle Ages. It was only after Tycho Brahe had demonstrated the impossibility of rigid celestial spheres that the comparison of the heavens with an airy or watery liquid and of stars with birds, fish or clouds had become once again prominent. With his view on this matter, Gorlaeus is found in good company with authors as diverse as Robert Bellarmine, Tycho Brahe or Johannes Kepler.[110]

The most probable source of Gorlaeus' acquaintance with up-to-date optical, mathematical and astronomical knowledge as well as cosmological issues is Adriaan Metius (1571-1635). In 1594, Metius had been one of Tycho Brahe's assistants on the Island of Hven. Four years later, he accepted a professorship at the university of Franeker, where Gorlaeus took his Arts degree. There, he taught mathematics, astronomy, surveying, navigation and fortification.[111] That the technical insights and theories that Metius had brought back from Tycho's island became common knowledge in the university of Franeker's scholarly community appears from the disputations held under the auspices of Henricus de Veno, professor of natural philosophy and ethics as well as Gorlaeus' teacher and landlord, which will be examined below.[112]

The most recent celestial astronomical discovery that Gorlaeus refers to in his two books has already been mentioned: it is his reference to the telescopic observations announced in Galileo's *Sidereus nuncius* of 1610. Surprisingly enough, however, he only mentions the optical resolution of the Milky Way into a myriad of stars, which proved its supralunary nature and disproved notions that it was a sublunary, meteorological phenomenon. By contrast, no word is said about Galileo's discovery of the earth-like surface of the moon or of Jupiter's four moons.[113] Nonetheless, despite its brevity and restraint, this reference constitutes the earliest known testimony to Galileo's impact on the intellectual life in the Netherlands.[114] It shows that Gorlaeus, despite his otherwise inconspicuous reliance on the results of the empirical sciences, took an interest in, and had access to, some of the most recent discoveries and theories, at least in the domain of astronomy.

At the same time, it must be mentioned that the *Exercitationes* does not show any trace of Copernicanism. Quite to the contrary, Gorlaeus explicitly combats this view in a specific section entitled "The terrestrial globe does not move." This section illustrates however his awareness of the chief arguments in favor and against a heliocentric cosmology. Particularly interesting is his paraphrase of the Copernicans' response to heliocentric passages contained in the Bible: "To the places in the Holy Scripture, they respond that the latter speaks about motion, not how it is in itself, but how it is conceived by us."[115] Which Copernicans he may have had in mind, he does not say – but the most fascinating possibility is of course that he knew of the treatise of Rheticus, whose manuscript was kept at the Martenastate, where Gorlaeus grew up, before it was published, together with Gorlaeus' own *Idea physicae*, in 1651.[116] Judging by the evidence assembled in Rienk Vermij's study of the reception of Copernicanism in the Dutch Republic, there were exceptionally few Copernicans to be encountered in the Netherlands in the days when Gorlaeus wrote his treatise, although there were a handful of humanists interested in the Tychonic system and a small number of others who examined, but rejected, Copernicus' system. Not even "the telescopic discoveries by Galileo" managed at first "to elicit a noteworthy reaction at the university," as Vermij shows.[117]

Gorlaeus' decision to dedicate a section to the discussion of the motion of the Earth is, in this respect, certainly remarkable. However, after deploring the absence of demonstrative proof on the part of the Copernicans, he lists the standard reasons for rejecting the assumption of the Earth's motion: (i) it contradicts the natural motion of the element earth, which is rectilinear; (ii) it is inconceivable that one should not feel such a motion; (iii) a stone thrown up into the air would have to land elsewhere and not descend to the same place.[118]

We may conclude, then, that Gorlaeus represents the conspicuous case of a natural philosopher who remained committed to a geocentric cosmology while at the same time combating the Aristotelian distinction between a superlunary and sublunary physics. Since he does not discuss the issue of the ordering of the planets, we have no basis for determining whether he inclined towards a Tychonic model (in which the Sun orbits around the Earth, but the other planets circle around the Sun). However, his embedding of a geocentric cosmology within a universal aerial fluid is compatible with Tycho, and may be explained by reference to Metius' teaching at the university of Franeker.

If we take a step back from the niceties of his cosmology, and look at his physical system in its entirety, the following picture emerges. His combination of an atomistic ontology with an Aristotelico-Galenic notion of the temperament of mixts, Cardano's two-element theory, Scaliger's minimist theory of mixture and some occasional cosmological and chemical ideas yields a theory that is not in all respects coherent in defining the nature of either the atom or mixture. Furthermore, we have alluded earlier to the theological implications of his doctrine. We have seen that the physical atom is only a sub-class of all *entia per se*, of which God is the most important representative; by the same token, mixtures represent only one example in the category of *entia per accidens*, of which humans are the most important example.

In fact, Gorlaeus expends considerable energy on furnishing a proof that man is an 'aggregate entity' made up of body and soul, which, though co-existing within one another, nevertheless never loses its specific essences – a theory that was to trigger the Utrecht Crisis thirty years later.[119] The theological reasons for which it might have been so important for Gorlaeus to separate soul and body will be discussed below in detail.[120] Let us just anticipate here that in the eyes of Gijsbert Voetius, the anti-Arminian theologian who combated this thesis as late as 1641, it was linked to the heterodox views of the Arminians and to their understanding of the relation between man and God.[121] The same holds true of Voetius' son Paulus, who in 1657 still dedicated pages to a refutation of this thesis.[122]

Theological concerns also explain certain aspects of the last part of Gorlaeus' *Exercitationes*, where it is attempted to show that the principle that "nothing comes of nothing" (*ex nihilo nihil*) is invalid. The rejection of this principle may at first sight be surprising. After all, Greek atomism has often been depicted as an ingenious answer to the Parmenidean challenge to explain natural change without having to postulate the continuous generation *ex nihilo* of new bodies. For Gorlaeus, by contrast, atomism helps to prove the exact opposite, namely that God is not only everywhere present, but that he is continuously involved in the creation of entities and substances:

> Each substance that is made, is produced directly by God, and whatever comes about is made by God, and is created out of no substance. [...] Therefore, whatever substance is made, is made by God, and what perishes, is reduced by God into nothingness: by the same token, whatever is made is made out of nothing.[123]

The aversion to the *ex nihilo nihil* and *in nihilum nihil* axioms was taken from the writings of the German philosopher Nicolaus Taurellus, as we shall document below. Why it was necessary for Gorlaeus' own system is, however, not entirely evident, nor how it is compatible with the entire doctrine concerning mixture (which, after all, does not come about *ex nihilo*). However, one of the principal objectives of this reasoning is evident enough. It is to argue that the human body is an aggregate of atoms incapable of producing a higher form. God therefore does not only create the human soul *ex nihilo* on the day of conception, but also the *anima vegetativa* of all living beings, including plants, and the *anima sentiens* of all animals, including man.[124] For Gorlaeus, this view entails that spiritual and material *entia* are ontologically independent of each other and mingle only accidentally. Given how strongly the Platonized hylemorphism of Latin Christian Aristotelianism had blended the material and the spiritual realms, Gorlaeus' intellectual choice in favor of atomism must also be viewed in the light of the possibilities it offered to the theologian of separating the soul from the body – a view that had appeared as promising to Taurellus before him as it did to a range of Cartesians after him.[125]

Both treatises by Gorlaeus conclude on a discussion of the soul. The shorter treatise ends with a defense of free will (*liberum arbitrium*) of both man and God. The *Exercitationes* ends, by contrast, in a different and fairly anomalous manner. One is indeed forced to conclude that Gorlaeus passed away before he had been able to finish the concluding part of his treatise, as he numbered the existing section as '1' although there are none that follow it. Moreover, this last *exercitatio* differs markedly in tone and style from everything that precedes it. Here, Gorlaeus is found to paraphrase "answers" (*responsiones*) he had "once given to passages in the Holy Writings"; that is, a theological disputation, regarding the "famous controversy whether the souls were produced by God or by the parents."[126] As a matter of fact, the authorities he quotes in this concluding section are not philosophical but exclusively scriptural, and the manner of exposition is the disputational one of addressing arguments *pro et contra*.

This disputational question on the origin of the intellective soul addresses a notorious issue that divided creationism from traducianism. The issue of whether the individual soul was created *ex nihilo* by God or was instead generated by being handed down by the parents *ex traduce* divided philosophers at the time, and often along markedly confessional lines. According to Joseph Freedman, Lutherans tended to stress "that souls are generated, while their Calvinist and Roman Catholic counterparts asserted that God creates souls out of nothing."[127] This picture, while generally valid, is however too tidy. With respect to the case in hand, it will be shown that the opposing camps in the Calvinist controversy over predestination, the ineluctability of sinfulness, grace or damnation, could not have viewed the question concerning the origin of the soul and the possible contribution of the parents in a consensual manner.

Even in Gorlaeus' unfinished discussion of this issue, one is tempted to discern a distinctly Arminian penchant. He rejects the arguments of both camps, which according to him rely on sterile and apodictic Aristotelian demonstrations that he compares to "pleasing pigs with mud."[128] Gorlaeus' own position lies somewhere in the middle: "The soul is generated by the parents out of nothing (*ex nihilo*), thanks to a singular act of God's concourse, in which he decorates the soul with a variety of gifts and determines the parents' general potency of generation towards this particular individual."[129] This concourse, he adds, takes place in the moment of conception, and not only in the case of humans, but also of animals.[130] With this position, he seems to insert himself in a tradition that goes back to Gasparo Contarini (1483-1542) and was handed down through Goclenius and Zanchi to Casmann, and which posited that the parents provided the indeterminate matter of the child-to-be, and God the form.[131]

While the issue of the origin of the soul belonged to the realm of a Christianized natural philosophy, Gorlaeus' theological embedding and thrust is evident not just from his discussion of scriptural passages, but from the question he addresses in the last seven pages of his treatise: it is the "urgent" question of original sin.[132] This problem is urgent because, on the one hand, sin "cannot be just in the body, as it is a blindness in the intellect and a perversity in the will, of which only the soul, not the body, is capable." On the other hand, it is also evident that God is the author of the soul. The question therefore evidently arises as to whether "God is the author of sin."[133]

Gorlaeus somewhat exuberantly claims that he knows how to resolve this age-old, ominous problem. His solution is both unconvincing and wild. In his attempt to drive a wedge between the Good God and the soul of fallen man, he proposes a doctrine that might anachronistically be labeled biological determinism. The evil innate in the human body and handed down from one generation to the next, he argues, is due to the individual ordering and disposition of the organs. Is it not known that people with a hot temper are, as it were, organically more disposed towards wrong-doing than others? "Many of the actions of the soul depend on this temperament. For they cannot be good, as they do their evil because of an organ."[134] The solution to the tricky problem of God's involvement in our wickedness is thus two-pronged: God bestows on the newly created soul all the original perfections that Adam had possessed before he had bitten into the apple, plus the possibility to deviate from them (what the scholastics, and Gorlaeus with them, call *privatio*). The parents, in turn, bestow on their child some positive organic "proclivity towards evil."[135]

Having settled this theological issue to his own satisfaction, Gorlaeus wishes to add something more general about the origin and disappearance of souls. After a detailed, almost phenomenological description of the act of sexual copulation, in which he argues there is no precise moment in which one might possibly claim that the soul was created, he turns to the issue of spontaneous generation, which he takes to provide yet another proof of the intervention of God. The worms in the cheese were not in the milk, nor in the cow that made it, but were planted there by divine intervention. Whether the vegetative soul also requires God's intervention is, by contrast, less obvious: branches, when cut off from the tree and planted elsewhere, are sometimes found to grow into new trees, as if the same soul were divisible. Do the three souls thus not behave analogously? Gorlaeus does not know, nor does he appear to have found the time to decide, as his own life's time was up. His *Exercitationes* end abruptly, in great haste, and on a very odd note: "What must be said about the vegetative soul is not yet clear to me. But what must be stated about the origin of the sentient and rational souls, namely that they are created by God, has already been demonstrated. End."[136]

The soul is the only subject matter on which the two books differ markedly. The concluding book of the *Exercitationes*, "On the soul," far from offering the natural philosophical analysis that one would have expected, seems to offer the mere paraphrase of a theological disputation whose focus lies on God's involvement in creating and defining the soul, and on the question of God's responsibility in handing down original sin. It is moreover characterized by a sinister emphasis on the "evil dispositions" that hamper the efforts of our will. By contrast, the concluding section "On the human soul" in Gorlaeus' *Idea physicae* not only provides a more worked-out theory of the soul, but concludes with a eulogy of free will.[137] Gorlaeus defines there the human soul as an immortal "spiritual substance," which does not need to rely on a bodily organ.[138] In an interesting argument, which in some sense anticipates a Cartesian idea, Gorlaeus adds that the soul takes most, but not all, of its truths from the senses; a number of "principles" are in fact innate, for example, that there is a God, or that there is a difference between 'one' and 'two'.[139] Turning to the issue of universals – a topic, we recall, that took up dozens of pages in the *Exercitationes* – Gorlaeus insists that the human intellect obtains universal notions by abstracting from singular ones (but evidently not the just-named innate 'principles').[140] After having discussed the intellect's modes of reasoning, he ends his treatise with the theologically sensitive issue of free will. Freedom, according to his definition, is the decision to "follow up or avoid the things that are understood in the intellect (*res intellectas*), by willing and choosing."[141] In contrast to the appetites of the senses, which desire only pleasure, free will relies on the intellect's prior assessment of whether an action is good, honest, pleasant and useful.[142] The action of the will is free – the contrary, an enforced will, would clearly not be a 'will' – and its freedom consists in its indifference, that is, its prior indeterminacy.[143] "The freedom of my will does not reside in my ability to will the good and not to will the bad, but in my ability to will (or not to will) a given object. [...] In this way God, who wills the good, wills it freely: for he cannot will the bad."[144]

With this shorthand version of a theodicy, Gorlaeus ends his discussion of free will and with it his entire treatise. The *Idea physicae* carries thus none of the more sinister deterministic overtones of the *Exercitationes*. One is left wondering, in fact, whether the *Idea*'s much more upbeat insistence on the freedom of will in its choice for the good comes not closer to Gorlaeus' real view of the matter, while the sinful determinism of the *Exercitationes* contains the traces of theological positions that had been dictated by the professor under whom he may have conducted the disputation to which the text refers.

2.5. A BRIEF APPRAISAL

Anyone who thinks of Gorlaeus as an early modern scientist will reach the conclusion that his philosophy is strangely suspended between a quantitative theory relying on extended but indivisible entities and a theory of qualities that appears to be incompatible with atomism. We have seen that according to his physics, most phenomena are caused by the 'real accidents' of warm and cold, which are not only continuously generated *ex nihilo* and annihilated *in nihilum*, but migrate freely from one entity to the other. Gorlaeus' world is indeed a far cry from both the materialism of ancient atomism and the mechanical models that became fashionable after Descartes in later decades of the century. His material atoms represent but one set of beings among others, which are immaterial. His is neither a mechanistic cosmos nor one in which final causes are abolished. In all these respects, Gorlaeus' model fails to conform to our notions of atomism ancient, early modern or modern.

Yet such a picture, while certainly not false, looks at Gorlaeus from the wrong perspective. Instead of pointing to impurities in his apparently half-hearted atomism, we should attempt to comprehend how he arrived at his particular type of atomism in the first place, why he needed it and how it resulted from his overall project, a project whose ontological nature we have been at pains to emphasize. Gorlaeus' is a world the main component of which is constituted by the *ens per se*, a category of 'unit of being' that counts as its members God, angels, souls, material atoms and real accidents. All of these beings possess unchanging essences. The being (*esse*) of God also includes a capacity to act (*posse*), which explains why God can create new beings. Almost all change in nature is due to the reintroduction of such new beings: some of them, such as the three types of souls (vegetative, sentient and rational), are singly created at their moment of conception; others, such as atoms, are endowed with 'forces' that propel them in God-given directions;

others again, such as the 'real accidents', spread through matter by multiplying themselves within various corporeal carriers.

Sure enough, this type of atomism may require a lengthy exposition of the type that was given above; but let us not conclude, for that reason, that Gorlaeus' physics is unnecessarily odd, awkward or bizarre. The sixteenth and seventeenth centuries' renewed sympathies for atomism nowhere resulted in a pure renaissance of ancient atomism. However much one may have praised Democritus as history's first physicist or have admired Lucretius' verse, there is no single so-called 'atomist' philosopher who would not have subscribed to Giordano Bruno's above-mentioned caveat: "atoms and void do not suffice." To begin with, no early modern atomist excluded God. It must however be obvious that the introduction of an omnipotent deity into an originally materialist world-view cannot possibly have left the ancient model intact. One finds, in fact, that early modern atomistic physics abounds in world souls, enlivening ethereal impulses, divinely implanted seeds or forces, sympathies and antipathies, God-programmed and therefore innate tendencies, and – with a renewed intensity after the triumph of Isaac Newton's physics – forces acting between bodies over indefinite distances. Viewed in this context, Gorlaeus' combination of atomism with other entities must not in the least scandalize us.[145]

This ontological messiness, as it were, also affects the motives behind the revival of atomism itself. It has been persuasively argued by Hans Kangro, Christoph Meinel and others that the empirical evidence that was adduced for the existence of atoms was for the most part illusory and rhetorical. Neither did seventeenth-century microscopes allow for their ocular inspection; nor was there a chemical proof for the existence of atoms.[146] William Newman has repeatedly insisted that reversible chemical processes and notably the 'reduction to the original state' (*reductio ad pristinum statum*) constituted a powerful proof of the atomic structure of matter.[147] His insistence is of course entirely justified. At the same time, it is also true that only few chemical reactions allow for this type of reversal. Moreover, the resulting quantitative type of atomism does not solve the problem of the appearance and disappearance of the emergent properties even in those chemical substances that are susceptible to such reversible operations. Why the mixing of non-toxic ingredients should yield a poison is not explained by a model in which inalterable and merely quantitatively or geometrically defined atoms intermingle into a contiguous, merely spatial vicinity. Moreover, a range of mutually incompatible atomistic models can explain the observed reversibility. In fact, an explanation à la Scaliger, Gorlaeus, Basson or Sennert, which proposes material units that are primarily defined in qualitative terms and whose properties allow for a modification in such a way that a certain 'union' comes about when they mix, is probably better suited to the purpose than the purely geometrical descriptions of atoms offered by those who attempted to emulate Descartes' mechanical model.[148]

But if it is true that empirical findings did not really necessitate an atomistic conception of nature, what else did? What were the motivations propelling the early modern champions of discrete material units? As a matter of fact, a plethora of alchemical, medical, natural-philosophical, ethical, historiographical and literary impulses can be discerned behind the reinvigorated interest in atoms.[149] Several of these exerted a direct or indirect influence on Gorlaeus, as we have adumbrated above and shall see in greater detail below. One additional motive must be mentioned here, which is usually ignored or contested, but which is quite forcefully present in the case of our young Dutch philosopher: theology.

It is usually held that theologians abhorred atomism. Sure enough, most Catholic and Protestant divines tended to view atomism as an outgrowth of Epicurean materialism and hedonism, decried it as a philosophy 'devised in brothels', and viewed its negation of substantial forms as a danger to the explanation of the Eucharist or the eternity of the soul. Yet, some philosophers came to prefer atomism precisely because of its theological potential: it allowed one to keep body and soul much more clearly apart than Aristotle's hylemorphist model, in which "the soul is the first grade of actuality of a natural body having life potentially in it."[150] Others liked atomism because it allowed one to abolish secondary causes and give major responsibility to the first cause, God, whose omnipotence seemed otherwise unduly diminished. Sébastien Basson, a Catholic and Jesuit-trained philosopher and physician who had converted to the Calvinist school of Protestantism, admiringly recalled that the Islamic school of atomism – the ninth and tenth-century Basrian Mutâzilī, who constituted the first large school of *kalām* – had delegated all power to God, who was responsible for recreating the world in each atom of time anew.[151] Basson had heard of this school through a commentary on Aristotle's *Physics* by the Jesuit Franciscus Toletus, who in turn based himself on Maimonides and Averroes. Toletus states:

> For it was the view of certain Arabs [...] who said that these inferior causes were ineffective, and that God alone produced the effect when they were present: hence God alone burns when fire is present and operates whenever any other agent is present. The particular causes themselves behave only as the signs of divine action. Among the theologians, Gabriel [Biel] accepted this [on the basis that] those things happen in vain through many causes that can happen through fewer; but God can carry out all things by himself alone. Hence the cooperation of particular causes is assumed in vain.[152]

Basson, who shared Gorlaeus' passion for this sort of radical usage of Ockham's razor, commented on this passage as follows: "These philosophers wanted, I say, the same as Plato, namely that God, being the sole principal cause, uses the others as instruments."[153] In his analysis of Basson, Lauge Olaf Nielsen has pointed out that wherever the desire can be discerned to attribute all activity to God, and to suppress secondary causality as much as possible, seventeenth-century atomism displayed generally Protestant or often specifically Calvinist overtones.[154]

Is this, then, also the ideological context through which we must explain Gorlaeus' metaphysical atomism? Or what else propelled this young philosophy graduate and first-year theology student to espouse atomist doctrines? In order to answer these questions, we must reconstruct his intellectual biography – which is precisely the task of our next chapter. We shall find, first of all, a series of natural philosophical and metaphysical ideas that Gorlaeus acquired during his studies at his first university, Franeker, and from his philosophy professor and landlord Henricus de Veno. It must have been there that he thought of writing his natural philosophical work, the *Idea physicae*. At the university of Franeker first and subsequently at the university of Leiden, where Gorlaeus enrolled in 1611 as a theology student, we shall moreover encounter an agitated religious environment in theological turmoil. These circumstances, we may surmise, led him to embed his physics in a new metaphysics, the result being the *Exercitationes*. Drawing our sums, we will be able to conclude that while a number of important textual sources contributed to his very precocious and often daring theories, the time and place of the composition of his two books, namely Franeker and Leiden in the years 1610-1612, suggests that the overruling motivation for developing his philosophy in the directions in which he did were theological.

Chapter 3

Gorlæus' Life

We recall from chapter 1 that Kurd Lasswitz, whose detailed analysis of Gorlaeus' natural philosophy of 1890 has provided the basis for all subsequent discussions, did not hide his perplexity at this author's idiosyncratic road to atomism. On the one hand, Gorlaeus' *Exercitationes philosophicae* (1620) were published in roughly the same years as the first edition of Daniel Sennert's *De chymicorum cum Aristotelicis consensu ac dissensu* (1619); the second edition of Nicholas Hill's *Philosophia epicurea* (1619); Francis Bacon's *Instauratio magna* (1620); Sébastien Basson's *Philosophia naturalis* (1621); Jean d'Espagnet's *Enchyridion physicae* (1623); and Galileo Galilei's *Saggiatore* (1623) – all works that contained some corpuscular or atomist concepts. Lasswitz therefore spoke of those years as a time in which "the corpuscular theory had already found wide diffusion and many supporters."[1] And yet, the peculiarly theological and ontological angle from which Gorlaeus arrived at his own positions looked sufficiently distinct from those chosen by his atomist companions for Lasswitz to decide that he could not view Gorlaeus as a member of an overarching consensus. However, unable to find any reliable information concerning the author's life and circumstances, he prefaced his textual analysis with that call for additional historical research that we have quoted above: "A monograph on Gorlaeus and on this important decade would be most desirable."[2]

A good part of the missing biography was unearthed by the Dutch professor of chemistry and historian Frans Maurits Jaeger, who in a seminal article of 1918 presented the results of his research into Gorlaeus' life and family background. Whereas Lasswitz had merely been able to discover that Gorlaeus had enrolled at Leiden's theological faculty in 1611, Jaeger managed to establish Gorlaeus' dates of birth and death, traced his paternal and maternal families, intuited their Arminian link and capped his achievement with his rediscovery of Gorlaeus' tomb under the wooden floor of the church of Cornjum.[3]

Paradoxically enough, Jaeger's numerous discoveries rendered Gorlaeus' philosophical works more mysterious, instead of explaining them. To begin with, knowledge of the exact date of his death permitted the dating of his writings to the period before April 1612; but that unexpectedly early date made our author appear even more original and atypical than Lasswitz had assumed. In addition, the discovery that these two works had been composed by a twenty-year-old student, who moreover was preparing for a career as a clergyman, rendered them even less comprehensible. As Gorlaeus' theories could now quite evidently no longer be interpreted as the product of bold abstractions from long scientific practice, they cried out for an additional explanation: what business did an aspiring theologian have with atomistic physics and metaphysics?

Jaeger's rich article, written in Dutch and published at the end of the First World War in a journal that enjoyed national circulation only, did not come to the attention of scholars outside the Netherlands. Remarkably, most Dutch historians also tended to ignore it, with the single exception of F.A.H. Peeters, who seventy years later republished the *Idea physicae* and had it translated into Dutch and Frisian – a publication that once more did not reach the wider world of learning. All of this explains why no one in a century has attempted to combine Lasswitz' presentation of Gorlaeus' philosophy with Jaeger's biographical findings, let alone search for a solution to the paradoxes that would have resulted from such a combination. Instead, as we have seen in chapter 1, twentieth-century historians of science have taken Gorlaeus to have been an experimental philosopher or even a chemist.

This chapter therefore proposes itself a task that could have been carried out a century ago: namely to complete the reconstruction of Gorlaeus' short life and to attempt an explanation of his views in the context of his personal circumstances.

3.1. GORLAEUS' FAMILY BACKGROUND

On his father's side, David Gorlaeus' family came from Antwerp. The family Van Goorle (also spelled as Van Gourle, Goirle, Ghoorle, Goerle, Gooirle, Gorle, and Gorlé, and with the Latin form Gorlaeus, used also in the present book) may have taken its name from the village of Goirle, which is situated in today's Dutch province of North Brabant. At any rate, the family is documented in Antwerp at least since the mid-fourteenth century, where its members were engaged in various trades.[4] David Gorlaeus descends from one Godevaert van Ghoorle Hendricxsone (1494-1558), a carpenter and cabinet maker, who had two sons, Peter and Jacob. Peter became a medical doctor and astrologer, who practiced in Antwerp and produced a range of Dutch and French prognostications from the year 1552 onwards, which proudly display the same coat of arms that we find on David Gorlaeus' tomb (see Figure 6).[5] Professionally speaking, his brother Jacob, the grandfather of our Gorlaeus, followed in his father's footsteps. When Jacob died in 1559, he left behind a number of children, among whom four by his second wife, Willemken Blockhuys – namely Sara, who would later marry a Dutch satin worker in Rotterdam, Abram; David (Sr.); and Govaert. We do not know whether and

FIG. 6: The simple version of the Van Goorle coat of arms (with a lion rampant as a primary charge, and a further lion as a crest) is given on the first page of Peter van Goorle, *Prognostication de lan [de] nostre Seigneur MDLXI* (Antwerp, 1561). The same coat of arms can still be recognized on David Gorlaeus' tomb (see Figure 1), although it has been brutally disfigured during the French period.

when the family converted to Protestantism. Clearly, David van Goorle and his brothers were Protestants when they entered early into the services the Prince of Orange, probably as financial administrators. It might have been for career reasons that they eventually chose to move from Antwerp to the Northern Netherlands, together with the epicenter of Protestantism and the princely court. We know, at any rate, that three of the four children of Willemken Blockhuys and Jacob van Goorle moved north: Abraham went first, David joined him later, and their sister, Sara, was the last to go.[6]

It must have been around 1580 that Abraham van Goorle (Antwerp, 1549-Delft, 1608), the uncle of our David Gorlaeus, and David van Goorle, Sr. (Antwerp, 1550-Cornjum, 1612), David Gorleaus' homonymous father, crossed over from what is now Belgium into the Netherlands (see Figures 7 and 8).[7] In 1582, Abraham had already left, as his brother David had to represent him in an Antwerp court case surrounding an illegitimate child. But soon thereafter, we find both brothers in the service of the Protestant nobleman Adolf van Nieuwenaar (c. 1545-1589). Van Nieuwenaar, whose family played an important role in the introduction of the Calvinist version of Protestantism in territories that nowadays lie on the German side of the Dutch border, had inherited from his father the county of Limburg and from his uncle the county of Meurs.[8] As a relative of Prince William the Silent, the leader of the Dutch rebellion against the Spanish, Adolf van Nieuwenaar was

FIG. 7: This portrait of 1580 by Hendrik Goltzius is usually said to depict Abraham van Goorle, but Jaeger, "Van Goorle," 224n1, convincingly argues that it shows his brother David van Goorle, the father of David Gorlaeus. That the man on the engraving is one of the Gorlaeus brothers is clear from the coat of arms (which is identical to that found in Figure 8, where the name is given). The general ledger (dated 1580) in which the gentleman depicted here is about to write is certainly much more compatible with David van Goorle's function as treasurer than with Abraham's functions. (Private collection)

a supporter of this insurrection. When he had to flee to the Netherlands after his defeat in a different conflict (the so-called Cologne Wars), the Dutch welcomed Van Nieuwenaar with open arms, electing him in 1584 to the stadtholderate – the highest executive office – of the Dutch Provinces of Gelderland and of Overijssel, and additionally in 1585 of the Province of Utrecht. At that point, Van Nieuwenaar established his headquarters in Utrecht's former episcopal palace.

From the extant documents from the 1580s, it appears that the brothers Van Goorle had become respectively counsellor and treasurer to the Count.[9] Because of these functions, they moved in 1585, together with the Count and the rest of the court, to Utrecht, where they settled at the Mariakerkhof.

In 1589, the Count died in a gunpowder incident which took place as he was inspecting new artillery pieces. His childless widow Walburgis (or Walborch) inherited her husband's possessions. The Van Goorle brothers now entered her services. In fact, Abraham van Goorle named one of his children after his new patron.

In 1591, David van Goorle got married. Utrecht's Protestant marriage registers contain the following entry:

FIG. 8: Portrait of Abraham van Goorle of 1601. Engraving by Jacob de Gheijn II. From Abraham Gorlaeus, *Dactyliotheca*. (Private collection)

> 3 January 1591. David van Goorle, treasurer to the honorable Lady Van Nieuwenaar, living in Utrecht; and Miss Sophia van Martena, living in The Hague. Married in The Hague.[10]

The exact date of the marriage in The Hague is unknown. However, only a few days separated their marriage from the birth of their son, David Gorlaeus, who was born on 15 January. That his birthplace was Utrecht, where his father – and obviously now also his mother – resided is testified by the place name *Ultrajectinus*, "of Utrecht," which accompanies his name in all official documents.[11]

The extreme vicinity of Gorlaeus' birth to the date of his parents' marriage is certainly surprising and allows for a number of obvious speculations. Perhaps this situation looked dubious even to Gorlaeus' contemporaries as according to an eighteenth-century document, he had to obtain a legal certificate in 1609/10 to prove that he was the legitimate child of his parents "through subsequent marriage" (*per subsequens matrimonium*) – although according to the official dates, he would not have been born out of wedlock.[12]

David Gorlaeus was three years old when his father was sacked in 1594 from his office on charges of fraud and abuse of office. A petition he submitted in December 1595 to the States-Provincial of Friesland and a letter sent by the latter to their colleagues in Utrecht allow us to conclude that Van Goorle was accused of having counterfeited the Count's signature back in 1588 in order to declare the Count's salary twice.[13] This allegation was, according to Van Goorle, false and the result of political machinations. His enemies in the government of the Province of Utrecht, as he furthermore reported in his petition, had in September 1594 confiscated all his possessions. When his wife, Sophia (or Swob, as she was called in her own tongue), had gone to Utrecht to have the family's belongings shipped to Leeuwarden, she was arrested while having lunch with her brother-in-law, Abraham van Goorle, and was only liberated because the Count of Solms (or possibly his wife) bailed her out. The Frisian States-Provincial accepted Van Goorle's version of the story. In their letter to their colleagues in Utrecht, they first explained why they felt called upon to defend Van Goorle, arguing that he was "a member of our province, because of the fact that he has married the eldest daughter of Doeke van Martena, one of the most eminent noblemen" of this province.[14] As for the issue at hand, the Frisians argued that Van Goorle had not committed any crime, but had instead been the victim of political intrigue.

Whether Gorlaeus' father had committed the crime that was ascribed to him or was removed under some false pretext for political reasons is hard to establish on the basis of the known documents. There are some historical circumstances that might point to a political trap. One has to remember that the rebellious Dutch provinces had in the 1580s tried to find a European monarch to protect and rule them before deciding, in 1588, to become a republic. One of the monarchs to whom the Dutch had in vain offered their sovereignty was the English Queen Elisabeth. After the assassination of Prince William of Orange and the ensuing turmoil, the English contacts intensified and resulted in a military treaty (the Treaty of Nonsuch). The arrival of troops in 1585 under the leadership of Robert Dudley, Earl of Leicester, led to a three-year, politically unsuccessful attempt to define the Dutch provinces as an English protectorate, and to institute the Earl of Leicester as the country's Governor-General. Queen Elizabeth did her best to undermine the success of this costly enterprise. A lack of political backing, Dutch disunity, Spanish military successes and mutinous English troops eventually led to the Earl's return to England in late 1587.[15] After his departure, factional fighting broke out in various places, including Utrecht, where Leicester had been based. In the end, Leicester's earlier supporters were removed from local positions of power. In the process, prominently placed Flemish immigrants were similarly eliminated. This process of purification may possibly have constituted the background to the downfall of Gorlaeus' father, who belonged to that latter group – as did the Count of Nieuwenaar himself, who had first supported, then dropped, the Earl.[16] This interpretation seems to receive further corroboration from the fact that David's brother, Abraham van Goorle, decided at the same moment to withdraw from politics, although we are not aware of any accusations that were levelled against him. Furthermore, it appears from documents that from 1597 to her death, the Countess Nieuwenaar paid both David and his brother Abraham a yearly allowance.[17] Even David's testament of 1612 mentions these payments.[18] This seems to suggest that David Sr. had been cleared of the allegations.

When David van Goorle lost his office, he and his family moved to Martenastate, the mansion of his father-in-law, Doeke van Martena, in Cornjum (see Figure 9). At this point, the life of little David and of his family assumed a Frisian character that was all linked to the Martena family.

David Gorlaeus' maternal grandfather, Doeke (also spelled as Doecke, Duco or Doco) van Martena (1527/30-1605) was a Frisian nobleman and courtier as well as the head of one of the province's most famous noble families, which resided in Cornjum at least since the early fifteenth century. In addition to the Cornjum estate, the family also owned city palaces in Franeker and Leeuwarden.

Doeke van Martena, who was destined to be the last scion of this family, became nationally renowned for the role he played in the Frisian and Dutch wars

FIG. 9: The stately mansion of the Martena family was burnt down in 1572 and subsequently rebuilt in 1584 by Gorlaeus' parents, Swob van Martena and David van Goorle, whose property it became. Between 1658 and 1899, when it was demolished, the mansion was reconstructed several times. This drawing by Jacobus Stellingwerf shows the Martenastate in c. 1725. (Courtesy of Fries Museum, Leeuwarden)

of independence, and went down in local history as the "Frisian Prince William," with reference to the leader of the Dutch revolt, William of Orange.[19] First a member of the States-Provincial and since 1564 of the States-Deputed (*Gedeputeerde Staten*) of Friesland, he early on and very openly sided with the Protestant party, and in due time chose to side with the Orange camp against the Spanish. In the turmoil of the ensuing wars of independence, he was forced to escape from his estate in 1572 when it was burned down by the pro-Spanish Frisian stadtholder Caspar de Robles. Named First Admiral of Friesland by Prince William of Orange in 1573, Van Martena took part in the naval battle of the Zuiderzee. After the so-called Pacification of Ghent in 1576, he could return to his lands, but was taken prisoner in 1580 by the troops of George of Lalaing, Count of Rennenberg, who had succeeded De Robles as stadtholder of the provinces of Friesland, Groningen, Drenthe and Overijssel and who tried for a last time to impose the rights of the Spanish crown on the population. Van Martena was released approximately one year later in a prisoner exchange.

Despite his continuing political role – he was named a member of the States-General – Van Martena was derided by his foes and praised by his friends for the fact that he had consumed "the fortune of three nobles" in the wars and died a poor man.[20] In fact, his poverty explains in part why Gorlaeus' parents moved to Cornjum. It appears that one of the tasks of David van Goorle was to assist his father-in-law, partly by managing the latter's finances. When Doeke van Martena died in 1605, his debt to his son-in-law amounted to 2560 florins; this is why the family mansion, the Martenastate, became the property of the Van Goorle family.[21]

We have also just heard that the Martenastate was burned down in 1572. As Doecke van Martena seems to have settled in his city residence in Leeuwarden, it might have been that when David Gorlaeus' parents left Utrecht to settle in Cornjum in 1594, they first had to reconstruct the castle. That the Martenastate could pass into the hands of the Van Goorles was also due to the fact that the Van Martena family was left without male descendants. Of the six children of Doeke van Martena and his second wife, Catharina Unema (or Trijn Oenema), all four sons died, childless, prior to their father's death. Doeke's own death on 11 November 1605 therefore marks the end of the Van Martena dynasty. For this reason, Swob van Martena, David Gorlaeus' mother and Doeke's oldest daughter, could take over *Martenastate*, once her husband, David van Goorle, Sr., had to leave Utrecht.

David Gorlaeus' father died in 1612, a few months after his son. His tomb, like that of his wife, who died in 1614, is located next to their son's in the local church.[22]

We must here also briefly return to David Gorlaeus' uncle, Abraham van Goorle. Quitting his function in the Nieuwenaar household at the same time as his brother, possibly because of the anti-Flemish sentiments that may also have swept

David van Goorle from his position, Abraham moved to Delft in 1595, where he lived as an independent and wealthy burger, dedicating himself to his ever more voluminous collection of coins and seals.[23] Although the two brothers shared a passion for the collecting of antiquities – a passion with which they may well have become infected in their hometown, Antwerp, where collecting was much more en vogue in those days – Abraham's own collection of coins, engraved gems and sea shells was to become internationally famous, drawing numerous visitors, including some from abroad. Both before and after his death in 1608, parts of his collections ended up in the possession of various European princes, including Henry IV of France, the Prince of Wales (Henry, the son of James I), and Queen Christina of Sweden. Abraham van Goorle, who was always on the outlook for potential buyers, documented his collections in a number of important numismatic works. The first, richly illustrated edition of the *Dactyliotheca* (documenting 400 engraved rings) was published for the first time in 1601, with numerous editions following throughout the seventeenth and into the eighteenth century. The author's embeddedness in humanistic circles, notably in Leiden, can be fathomed by the names of the authors of the poems that introduce his book: they are Hugo de Groot (Grotius), Joseph Justus Scaliger, Janus Dousa and Daniel Heinsius. His *Thesaurus numismatum Romanorum* (illustrating 450 Roman coins) appeared posthumously, in 1609, with a preface by his homonymous son and with a dedication to King Henry IV of France, who may have been viewed as a potential client.[24] The posthumous *Paralipomena*, in turn, is said to have been edited by the theologian and philosopher Petrus Bertius, who was a friend of Abraham van Goorle.[25]

3.2. GORLAEUS' YOUTH

These, then, were the prestigious and well-to-do circumstances into which David Gorlaeus was born. His parents having moved to Friesland in 1594 when he was three years old, David Gorlaeus grew up at his grandparent's rebuilt mansion in the small village of Cornjum, which even today counts less than 500 inhabitants.

He is likely to have first attended Cornjum's village school. That such a school existed, that it was supervised by the local parson and that David's grandfather and father paid for its costs is evidenced by a number of documents from the first years of the seventeenth century.[26] In the church of Cornjum, where David Gorlaeus is buried, we may in fact still see the testimonies of his parents' role in the village: on the walls hang two wooden panels, dated 1602 and 1608 and carrying the coats of arms of David van Goorle, Sr., and Swob van Martena. The earlier of the two reproduce the Ten Commandments; the later the Creed and the Lord's Prayer (see Figure 10).[27]

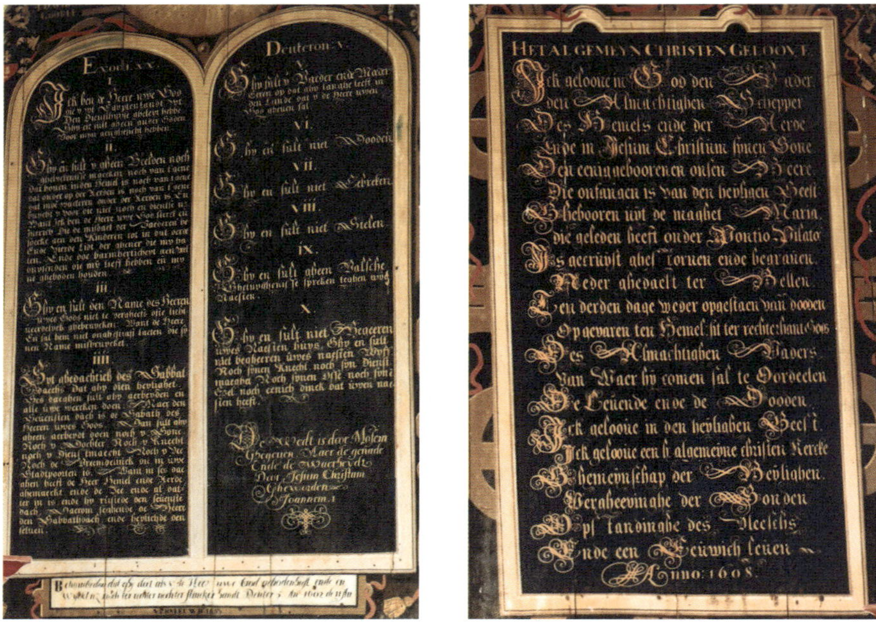

FIG. 10a and b: Even today, the walls of the church of Cornjum record the presence of Gorlaeus' parents as the village's seigneurial family. Two wooden panels, dated 1602 and 1608 and brandishing the names of David van Goorle and Swob van Martena, carry the Ten Commandments (10a) and the Creed (10b). (Courtesy of Fotoarchief, Instituut voor Christelijk Erfgoed, Rijksuniversiteit Groningen)

We must assume that Gorlaeus subsequently attended Leeuwarden's Latin school, which after the Reformation had been established in the Keimpmanastins, an aristocratic town residence that today counts as Leeuwarden's oldest existing house. Given that in the summer of 1606, Gorlaeus enrolled at the University of Franeker, we may assume that he spent the years prior to that date at the Latin School, which lay at a distance of six kilometers – a brisk hour's walk – from Cornjum's Martenastate. Whether he moved in with his grandparents, who then resided in the Frisian capital, or walked up and down, is not known.

In the period that Gorlaeus presumably went to Latin school, its rector was a certain Johannes Fungerius (Fungeri or Funger, 1546-1612), who had previously studied medicine and law in Leuven and Cologne. A prolific author of didactic poems, he seems also to have been interested in pedagogy, as is testified by the title of his "Book on the Discipline and Right Education of Boys" of 1586.[28] It should in this context be mentioned that in 1607, the year after Gorlaeus had left the Latin School, a certain Edo Neuhusius (Neuhaus or Nijenhuis, 1581-1638) was appointed as Fungerius' successor as rector. Neuhusius was born in the nearby German town of Steinfurt, the famous Academy of which we will encounter repeatedly below. In fact, his uncle, who educated him after the death of his own parents, was the well-known philosopher Otto Casmann, whose views Gorlaeus would have encountered in the lessons of his Franeker professor De Veno. Neuhusius furthermore counted among his teachers at Steinfurt the famous philosopher Clemens Timpler, whom we will also encounter below, as well as the theologian Conrad Vorstius, to whom this book will dedicate quite a number of pages.[29] The relevance of this link between the Steinfurt Academy, Frisia and the Arminian faction of Dutch Protestantism will soon become evident.

When his famous grandfather passed away in 1605, David was 13 years old and just about to enroll at Franeker University. When his apparently equally impressive and belligerent grandmother died three years later, David was 16 years old and about to obtain his first university degree. The circumstance that Gorlaeus passed the years of his youth in the presence of his aristocratic and battle-proven grandparents does not just add some touches of color to his biography, but also sheds some light on David's behavior. A number of Gorlaeus' own character traits might have been influenced by the unpliable and heroic character of the couple in whose mansion he grew up and who must have had a hand in his education. Take, for example, the stormy motto that Gorlaeus penned into a friend's album in 1610:

> Even if everyone holds the opposite view, the truth is to be defended; and one has to judge one's teacher in the same way as one judges one's greatest enemy. Keep this as your perpetual rule.[30]

The air of self-confidence, rebellious spirit and love of independence that these lines exude may owe much to the spirit of pride and defiance that must have reigned at the mansion in Cornjum and the town house in Leeuwarden, whose owners had accepted temporal exile, imprisonment and a lasting loss of property and wealth in the Dutch wars of independence.

3.3. GORLAEUS AT FRANEKER UNIVERSITY

As we have just heard, Gorlaeus enrolled at Franeker University in 1606. He was 14 years old, a normal age in those days for beginning one's Arts studies. The Frisian university of Franeker (see Figure 11), which lies at a distance of 20 kilometers from both Cornjum and Leeuwarden, was founded in 1585 as the Dutch Republic's second university, after Leiden (1575). A quite unique feature of the new Frisian university was that, in contrast to other early modern Dutch universities, its statues did not prescribe the teaching of Aristotelian philosophy.[31] The only non-negotiable requirement for its teachers was that they regard themselves as an integral part of the Reformed Church and did not violate the doctrines of the Hei-

FIG. 11: Franeker University looked very much like this when Gorlaeus studied there. This (hand-colored) illustration is taken from Pierius Winsemius, *Chronique ofte historische geschiedenisse van Vrieslant* (Franeker, 1622). Incidentally, Winsemius was also one of De Veno's students. (Courtesy of Tresoar, Frysk Histoarysk en Letterkundich Sintrum, Leeuwarden)

delberg Catechism. Indeed, Franeker's professors of theology made sure everyone understood the link between theology and the other disciplines.[32] In Frisia, the Reformation had gained the upper hand only recently, in 1580, in a very fast and violent operation, and the establishment of the university was intended to produce the new Protestant elite that the province needed. Philosophy, which was viewed as subordinate to theology, was expected to contribute to this enterprise, but divergent views quickly developed as to how this contribution could best be defined. Rivalling proposals as to how to reconcile philosophy with Protestant theology were offered, and not all of them relied on the Aristotelian corpus. While Lollius Adama (1544-1609), for example, still explicitly followed in the "footsteps of the Preceptor" (Aristotle), in 1610, the Ramist logician Frederic Stellingwerff (d. 1623) called Aristotle dismissively "that pope of nebulous opinions." [33]

It was at this nascent and still rather small institution that Gorlaeus enrolled in 1606, as the university's 928th student, signing up for philosophy, the discipline in which he presumably took his Arts degree three or four years later.[34]

In his work on the philosophical teaching at Franeker, Sybrand Galama has singled out two anti-Aristotelian figures whom he assumes to have been of particular influence on Gorlaeus' intellectual development, namely the young law student

Frederic Stellingwerff (d. 1623) and the professor of philosophy Henricus de Veno (c. 1570-1613).[35] Both figures have justly been associated with Gorlaeus in the subsequent scholarship, although partly for the wrong reasons. As we shall now see, these two men played very different roles in Gorlaeus' life. About Stellingwerff we may be brief, whereas De Veno will deserve a very detailed treatment.

Frederic Stellingwerff, who was a few years older than Gorlaeus, studied law.[36] The two young men appear to have known each other even before Gorlaeus moved to Franeker, because they showed up together at the auction of Alardus Auletius' vast library in June 1606.[37] Stellingwerff, who stayed at the house of Franeker's mayor, Hobbe Jelles Ansta, seems to have been well connected and may even have been appointed as mentor to the fourteen-year old Gorlaeus. In a rhetorically self-deprecating gesture, Gorlaeus signed off a touchingly adolescent panegyric on Stellingwerff, which was attached to a printed disputation the latter held in 1609, with the words: "David Gorlaeus of Utrecht, yours eternally, the one whom you have come to know so intimately, has hissed (*stridebat*) [this poem]."[38] However, while Gorlaeus may have admired the older Stellingwerff, the latter seems to have relied on Gorlaeus, who was of a higher social standing, for moral and quite probably for financial support. In his logical disputations, which he published in 1610, Stellingwerff publicly acknowledges this debt:

> Here, reader, you have my scholastic disputation which I elaborated about two years ago at the Frisian Academy of Franeker, when I gave private lessons, and which are now published at the instigation of the young David Gorlaeus, my intimate friend, who himself is setting in motion much bigger things than these.[39]

The last words may well contain a reference to Gorlaeus' own work-in-progress – probably the *Idea physicae*, or possibly already the *Exercitationes* – which he would have discussed with his friend Stellingwerff. Intellectually, however, the two young men went quite separate ways. While Gorlaeus tried to cast the foundations of philosophy anew by means of his ontology, Stellingwerff followed Ramus and chose dialectics. Although the above-named auction catalogue of 1606 evinces Gorlaeus' initial interest in Ramus – he bought a number of Ramist treatises, including an expensive compendium containing a *dialectica Rami* and an *Arithmetica Rami*, as well as Johannes Piscator's *Animadversiones in dialecticam Petri Rami* of 1580 – Gorlaeus' extant treatises betray no discernible debt to either Ramus or Stellingwerff, nor do the latter's later publications indicate any influence by Gorlaeus.[40] What the two shared was at any rate a rebellious dissatisfaction with traditional school philosophy, from which they both attempted to break away in their different ways.

Gorlaeus' intellectual debt to Henricus de Veno, professor in philosophy at Franeker, was undoubtedly more substantial. Very importantly, this influence did not just take place through classroom teaching. For, as the fascinating 1606 auction catalogue of the Franeker sale of the library of Auletius also reveals, Gorlaeus lodged at De Veno's house.[41] This sheds a very new and intriguing light on the student's debt to his teacher. Given that De Veno hid behind his façade quite a few personal secrets and probably nurtured a number of intellectual heresies, which we shall now have to examine, we may presume that Gorlaeus obtained from him far more unorthodox ideas than can be documented on the basis of written records.

3.4. HENRICUS DE VENO'S SECRETS

The figure of De Veno (see Figure 12) has fallen into oblivion, even among historians of Dutch philosophy and science, because he did not publish any works under his own name.[42] However, various European libraries contain published disputations that accompanied his philosophy courses, and from these disputations, De Veno emerges as a most unusual teacher. He may fairly be described as

FIG. 12: Henricus de Veno. Portrait by an anonymous painter, originally hung in the Senate Chamber of the Academy of Franeker. The portrait is analyzed in Ekkart, *Franeker professorenportretten*, 74-75. (Courtesy of Stichting Museum 't Coopmanshûs, Franeker)

the least orthodox and most original Dutch professor of natural philosophy of the opening decade of the seventeenth century. His philosophical approach was theologically grounded and at the same time heavily indebted to the Italian naturalism of Girolamo Cardano (1501-1576) and Julius Caesar Scaliger (1484-1558). Although Gorlaeus was to outrival him in productivity, coherence and intellectual force, De Veno's unorthodox views must be viewed as the necessary precondition for Gorlaeus' own metaphysics and physics.

While the philosophical views that De Veno taught during his lecture courses can in part be reconstructed on the basis of the extant disputations, his private opinions and convictions cannot. That is all the more deplorable as there exist forceful reasons for assuming that De Veno had, as it were, also an esoteric doctrine that differed from his exoteric, public teachings. These reasons emerge from his biography, which contained a dark secret that no one seems to have known at Franeker. One may assume that Gorlaeus, who lived in his house and will have overheard him converse at dinner, must have seen and absorbed more of his audacious ideas than other students.

Let us therefore first turn to the man himself. Henricus de Veno (who also spelled his name De Veen and Van der Veen) was born in the Frisian capital of Leeuwarden around 1574, as the second son of Jantje Gerrits Mamminga and of Laurents de Veno, secretary of Leeuwarden's city council and town magistrate. His three brothers were to obtain influential positions in the army and trade courts, while his sister married Johannes Rhala, the administrator of religious properties in Frisia (*ontvanger van de geestelijke goederen*).[43] Given these positions, we may presume that the De Veno family was acquainted with Gorlaeus' grandfather, the politician and army leader Doeke van Martena, who after the destruction of his mansion in Cornjum in 1572 resided mostly in Leeuwarden.

Having also graduated from Leeuwarden's Latin School, de Veno enrolled at the University of Franeker in 1591 as a student of "philosophy, languages, and theology."[44] Franeker University was at the time still an extremely small institution with an uncertain future. De Veno was in fact the university's 130th student and only one out of 18 to enroll in the year 1591. Maybe for that reason, he soon thereafter moved on to Leiden University, where he was awarded a master's degree in philosophy in August 1593, after successfully defending a number of intellectually inconspicuous logical and physical theses under professor Antonius Trutius, one of those early Dutch professors "whose names are not found in the history books."[45] In 1595, De Veno reappeared in Franeker, still as a simple *magister*, and in 1596 defended a set of theological propositions on usury there.

Instead of completing his theological studies, as one might have expected, De Veno next embarked on a most adventurous academic pilgrimage. Such tours usually took Frisian students to leading Protestant universities such as Heidelberg, Marburg, Basel or Geneva, where they would try to obtain their higher degrees.[46]

When De Veno returned to Frisia in early 1599, he claimed to have done just that: he declared himself to have become a triple doctor in law, medicine and philosophy, having also obtained much expertise in theology (albeit without possessing a doctorate in that discipline). He would henceforth sign with his three doctoral titles and did not prevent students from calling him "thrice great" for these qualifications.[47]

Yet behind this impressive façade lured an embarrassing and potentially dangerous secret. De Veno's collection of titles was, at least in part, his own invention – and this invention served to cover up a dishonourable and dangerous fact: De Veno had gone to Rome, the capital of the confessional enemy, and had there been denounced, arrested and imprisoned by the Inquisition! The dates of his arrival in Rome and of his arrest are not known with precision, nor are the exact charges that were brought against him; his specific file (like hundreds of others) was lost in the period when, on Napoleon's orders, the Archive of the Holy Office was shipped to Paris and only partially carried back to Rome after Napoleon's defeat. However, what emerges from the extant acts of the Congregation of the Holy Office is that by 3 June 1597, De Veno was in prison after having been denounced by a Scotsman called Robert Brown.[48] The Inquisition was supposed to have jurisdiction over all baptized Christians, including Protestants. As a Protestant, De Veno was considered a heretic, and heresy was viewed as a severe crime on a par with high treason (*crimen laesae maiestatis*).[49] Whether he had committed any more active 'crime', such as trying to spread Protestant ideas or texts, is not known, but would not have been necessary for his arrest.

De Veno, when brought to trial in the autumn of the same year, did confess that he had embraced Protestant heresies until the age of eighteen, but at the same time insisted that he had relinquished his heretical views by the time he was twenty-three years of age.[50] Although, as we have seen, De Veno had the year before still been studying theology at Franeker, he apparently tried to persuade the papal magistrates that between 1591, when he had first enrolled at Franeker, and 1596, when he left Friesland, he had gradually lost his Protestant faith, and that by the time he entered the Italian territory he had formally converted to Catholicism. It is not surprising that this account neither convinced the cardinals of the inquisitorial court, nor Robert Bellarmine, who was at the time a simple *consultor* working on this case. They regarded De Veno's statement at least as a partial confession and decided to have priests from the Low Countries visit De Veno in prison to obtain the whole truth. In March 1598, they sent the well-known Flemish theologian and editor of patristic works Gerard Vossius (1540-1609) to De Veno's cell so as to bring the latter to a full confession.[51] It seems that these visits produced at least some of the desired results, because in June 1598, the cardinals reached the verdict that De Veno had to abjure as a "formal heretic," which meant that his heresy had been proven. By abjuring, he returned officially to the Catholic faith.[52]

Usually, such a sentence would have resulted in a rather long prison term. It is therefore surprising to read in the acts that De Veno was released from prison within less than a week of this verdict. Although he was not yet allowed to leave Rome, he was granted an allowance for living expenses. Even more unexpectedly, in September 1598 he was given permission to return to his native Frisia.[53] The documents suggest that the Inquisition's lenience may be explained by De Veno's young age, his education by Protestant parents and his foreign provenance, which were all regarded as mitigating circumstances. The Inquisition was incomparably more severe towards Catholics, and notably Catholic Italians. At the same time that De Veno stood trial, Tommaso Campanella (1568-1639) was jailed for several years, tortured, and eventually confined to Roman and Calabrian convents. Even more notorious is the trial of Giordano Bruno (1548-1600), who after a seven-year trial (1593-1600) and a fair amount of torture was burnt at the stake, having refused to abjure his heresies.

As for Giordano Bruno, it is for the intellectual historian suggestive in the extreme to discover that De Veno was confined to the same prison as this famous heretical atomist. There exist at least two separate lists of prisoners visited by the inquisitors that name Bruno and De Veno side by side (see Figure 13).[54] As the number of prisoners was small – thirteen in one case and twenty in the other – it was inevitable for the two men to have encountered one another, although no such encounter is recorded in the extensive Bruno scholarship. But unless De Veno had denounced Bruno from within the prison for some heretical statements he had overheard (as other prisoners readily did), there was no reason for their possible conversations to result in any written record. With regard to Gorlaeus' atomism, written down only a decade later, it is highly tempting to look for a possible Brunian influence: the idea that heterodox Roman prison conversations were repeated during equally heterodox dinner conversations at Franeker and sedimented in Gorlaeus' daring treatises is highly alluring.

Unfortunately, however, in De Veno's disputations defended at Franeker, there is no discernible trace of any sympathy for atomism. De Veno's scepticism vis-à-vis Aristotelian natural philosophy or his emphasis on primary, divine causation are much more easily explained through his reading of Cardano, who is acknowledged as a source, than through Bruno. However, we do not know exactly what De Veno taught in his lecture course, and even less what he talked about at the dinner table with his precocious student.

But what had motivated De Veno to visit Rome in the first place? Had he really converted to Catholicism, as he affirmed in his court hearings? If not – as the course of his trial suggests – what attracted this Calvinist theology student to Rome? We do not know. But his case may be similar to that of the Dutch theologian Jacob Arminius (whom we will encounter repeatedly below), who as a student

FIG. 13 a and b: Record of the visit of seven cardinals of the Holy Office to the prison of the Inquisition, on 16 March 1598. Although the pages are ink-corroded and therefore difficult to reproduce on a photograph, one recognizes (Figure 13a, bottom) an inscription beginning with "Frater Jordanus [...] Bruni de Nola" (Giordano Bruno). On the facing page (Figure 13b, l. 14), another entry describes the subsequent visit to "Henricus de Veno de Leovardia." Decree of the Congregation of the Holy Office, ACDF, SO, *Decreta, 1598*, fols. 241v, 242r. (Photographs by Leen Spruit)

had visited Padua and Rome. "In later years," the historian of Arminianism Archibald Harrison reports, "it was asserted by his enemies that he kissed the pope's toe in the eternal city, formed an acquaintance with Cardinal Bellarmine, came under the influence of the Jesuits and secretly renounced the reformed religion."[55] What in the case of Arminius was mere calumny was however quite true for De Veno: he had met Bellarmine during his trial, and he had openly abjured his Protestant creed. It is obvious why he preferred to hide this ignominious episode from his fellow citizens back at Franeker.

At any rate, on 9 September 1598, De Veno left Rome. He did not tarry and speedily removed himself back to Protestant lands. Two months later, we find him enrolled at Basel University, where an entry in the *Matricula* of the Theological Faculty states (see Figure 14):

> Henricus de Veno, Frisian. Declares that after becoming doctor of law in France, he furthermore wished to finish his study of theology. He was detained for an entire year in Rome in the prison of the Inquisition.[56]

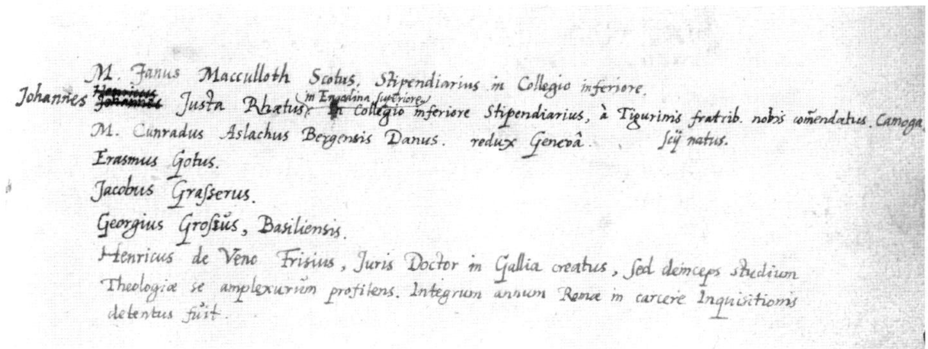
M. Janus Macculloth Scotus, Stipendiarius in Collegio inferiore.
Johannes Justa Rhætus in Engadina superiore in Collegio inferiore Stipendiarius, à Tigurinis fratrib. nobis comendatus. Camoga
M. Cunradus Aslachus Bergensis Danus. redux Genevâ. Sey natus.
Erasmus Gotus.
Jacobus Grasserus.
Georgius Grossius, Basiliensis.
Henricus de Veno Frisius, Juris Doctor in Gallia creatus, sed deinceps studium Theologiæ se amplexurum profitens. Integrum annum Romæ in carcere Inquisitionis detentus fuit.

FIG. 14: De Veno's entry in Basel's *Matricula facultatis theologiae,* 1597-, p. 43, in which he declares his previous academic activity and his imprisonment in Rome. (Courtesy of the Universitätsbibliothek Basel)

De Veno's frank admission that he had wished to pursue his theological studies at the center of Catholicism may be surprising. It might indicate, however, that to Basel's university authorities, he suggested that he had gone there as a missionary for his own faith, and had ended as a martyr. Whatever the case may have been, what he admitted in Basel, he kept to himself back in Friesland. After all, studying in Rome was expressly forbidden to Dutch students by the government, and in fact no other Frisian Protestant is known to have tried to study theology there.[57]

In late 1598, then, De Veno had still not completed his theological studies, but claimed to possess at least a French doctorate in law. When and where he obtained this degree is unclear. However, De Veno was from a family of lawyers, practiced law for two years after returning home, and identified himself as a "doctor of law" already to the Roman inquisitors. We must not dismiss the idea that he had done sufficient coursework for a doctorate in law, either between 1593 and 1596, when he resurfaced at Franeker or between his theological disputation at Franeker in 1596 and his arrest in Rome in 1597.

However, De Veno did not stay long in Switzerland, nor did he complete his theological education there. Five months later, in April 1599, we find him practicing as a lawyer (*advocaat*) in his hometown, the city of Leeuwarden. In 1601, he applied for the position of professor of theology at his *alma mater*, the University of Franeker, but given that he had no theological diploma to his name, the Senate preferred to elect the French theologian Franciscus Junius (du Jon, 1545-1602) in his stead.[58] However, in October of the same year, De Veno was proposed for a new chair in ethics and physics. On 23 September 1602, his nomination was confirmed by the *Gedeputeerde Staten* and he began his career as Franeker's *Professor ethices et physices* at a salary of 600 florins per year.[59]

De Veno remained in his chair until his early death on 22 April 1613. As a teacher, he appears to have been quite popular among the students and was later fondly recalled by some of them.[60] However, historians of Franeker University record a grave incident that occurred in 1609, and of which Gorlaeus must have been a witness. After having assumed the rectorate of the university in June of that year, De Veno became involved in serious litigation with a number of his colleagues. The professors Marcus Lycklama, Timaeus Faber, Lollius Adama and his son Augustinus Adama, Adriaan Metius and Sixtus Arcerius collectively denounced him to the provincial government. He was subsequently suspended from his two charges as rector and professor, though at the same time the salaries of Augustinus Adama, Metius and Arcerius were lowered each by 100 florins as a punishment for their litigiousness.[61]

It is hard to reconstruct what may have triggered this scandal. Vriemoet suggested that De Veno may have overestimated himself and his universal competence and that his arrogance may have angered his colleagues. Given his biography and partially false pretences, this sounds plausible enough. By contrast, that his Roman imprisonment and (temporal) conversion to Catholicism had been discovered is not plausible, because the resulting scandal would have left traces in the Frisian historiography. At the same time, the occasion for the clash must have been more concrete than mere 'arrogance'. Vriemoet specifically surmises that De Veno's former teacher, the Aristotelian Lollius Adama, may have taken exception at the novel Platonist theses taught by his pupil.[62] Most subsequent historians, accepting this interpretation, speak of "battles between supporters and opponents of Aristotelianism."[63] But given that in 1609, De Veno had been teaching his peculiar philosophy for more than seven years, this seems an unlikely explanation.

It is much more probable to assume that De Veno's removal had a theological component. A few months before the row broke out, a battle that had originated at Leiden over questions regarding predestination but had meanwhile assumed much more important and menacing dimensions, had reached Franeker. Below, we will hear a lot more about this battle between Arminians (also called 'Remonstrants' after 1610), on the one hand, and anti-Arminians (in early stages also called 'Gomarists' and after 1610 'anti-Remonstrants'), on the other. Suffice it here to take note of the fact that De Veno's colleague, the theologian Sijbrand Lubbert (c. 1555-1625) had since 1604 been very much involved in it. "The north-easterly provinces of Friesland and Groningen were the stoutest supporters of High Calvinism in the Netherlands," according to Harrison, who adds: "In this zealous allegiance the University of Franeker led the way, and the mouthpiece of the University was Sibrandus Lubbertus."[64] Always on the lookout for unorthodox views, Lubbert had earlier started a controversy with Johannes Drusius (1550-1616), professor of oriental languages, whom he accused of inclining to the Arian heresy. In 1615, he would also vie against a further colleague of his, the theologian Johannes Maccovius (1588-1644), over what became known as the *causa particularis Frisica*, a controversy between supra- and infralapsarianism.[65] As Van der Woude writes in his biography of Sijbrand Lubbert:

> In all these years, he was engaged in fights on all sides. His campaign against Vorstius [Arminius' successor at Leiden] had not yet finished when the conflict with Drusius started and he had to defend himself against Grotius. The battle raged inside the sphere of Dutch Protestantism, nay, within the very walls of the Franeker Academy.[66]

Debates had flared up at Franeker violently during the first weeks of De Veno's rectorate – when Gorlaeus was still his lodger – and reached its first peak around the time when he was forced to resign as rector and professor. In June of 1609, Simon Episcopius (1583-1644), Arminius' talented student, who was later to become professor of theology at Leiden and one of the two main spokesmen for the Remonstrant party, showed up at Franeker to study with the professor of Hebrew, Drusius. Against all better advice, he had even allowed himself to get entangled in public disputations with Lubbert. It has in fact been stipulated that he did so on purpose, so as to weaken the reputation and influence of Franeker's self-appointed watchdog of Calvinist orthodoxy.[67] It has also been reported that Lubbert protested with the university curators about the Arminian faction within his own university.[68] Although none of our sources mention De Veno in this context, the temporal coincidence is striking enough. Beginning in 1609, Lubbert, who had for years been styling himself as a kind of anti-Arminius, made sure that the orthodox view retained the upper hand. As the confrontation turned from one between single theologians into one involving theological as well as political schools of thought, it also became more menacing and the measures taken more drastic. A new peak was reached in the winter of 1610-1611, when several theology students who had previously studied at the Steinfurt Academy with Arminius' appointed successor, Conrad Vorstius, had to flee from Franeker when the town and university authorities discovered that they were behind the publication of Fausto Sozzini's highly heretical *De officio hominis Christiani*.[69] The personal link between Gorlaeus and the students who published that explosive text will be analyzed below.

As for De Veno, there are various reasons that allow us to presume that in this ever growing doctrinal affair, he was leaning towards Arminius and his proposed successor, Vorstius. For example, his agreement with Jean Bodin's call for a strong government watching over a state of confessional tolerance was shared by Vorstius, who in 1610 called upon the Dutch States-General to keep the churches under tight control while guaranteeing a *libertas conscientiae*, a *nativa libertas* in doctrinal interpretation and a *prophetandi libertas* in expounding such interpretations publicly.[70] Quite generally, the liberties De Veno took in cooking up his own philosophy, rather than following in Aristotle's footsteps as most of his Dutch colleagues did, suggest precisely the kind of 'latitudinarian' and non-dogmatic attitude that Lubbert perceived as a public danger. Moreover, like other Arminians, he may even have hoped for an eventual reconciliation of the divided Christian confessions – a hope that may possibly explain his imprudent visit to Rome a decade earlier.

Arminius' friend, the eminent theologian Johannes Wtenbogaert, wrote as early as 1606 that doctrinal dissent was not dangerous for the Church, but might, if left to itself, eventually result in a greater consensus.[71] A perusal of De Veno's metaphysical, physical and political convictions, at least in so far as they have been preserved in his published disputations, would suggest that this Franeker professor might also have subscribed to this view. His colleague Lubbert, in keeping with his convictions, of course reacted negatively to all calls for greater liberty in theological matters, arguing that this implied opening the doors to heresy.[72]

In sum, then, once the Arminian issue had exploded at Franeker, it seems not implausible to assume that given his own theological and philosophical proclivities, De Veno did not take the steps and measures that Lubbert and the university's curators expected of him as the university's rector, or possibly even stepped beyond his duties in siding with Episcopius. Although Lubbert does not figure on the list of professors who denounced him, the temporal coincidence with the Episcopius incident is such that it is counterintuitive to assume that it played no role in his suspension as rector and professor.

Pointing to Frederic Stellingwerff's hasty publication of his Ramist dialectics in 1610, Klaas van Berkel has wondered whether this young philosopher might have hoped to replace De Veno.[73] If that was indeed Stellingwerff's hope, he must have been sorely deluded. De Veno was reinstated in his old chair on 28 January 1611, at the lower salary of 500 florins, on condition that he would from now on respect the laws of the university, "abstain from subtle *parerga* and *quaestiones*, and also from defamatory acts and words" and, finally, that he would stop his extracurricular contact with students, who seem to have sided with him during the clash with his colleagues.[74] It is quite possible to interpret these conditions for readmission as aiming at De Veno's theological extrapolations from strictly philosophical matters. Now, if there was anyone who had "extracurricular contact" with De Veno, it was Gorlaeus, who lodged in his house!

It is a shame that we do not have any printed disputations held under De Veno from this turbulent period. We are therefore in the dark with respect to the contents and possible evolution of his philosophical, theological and political views in the years in which Gorlaeus studied with him. What we do know is that after his reinstatement, De Veno taught for two more years. He died prematurely on 22 April 1613, at roughly forty years of age.

3.5. HENRICUS DE VENO'S TEACHING

As most of De Veno's extant disputations are kept in libraries outside the Netherlands, historians of Dutch philosophy or of Franeker University have not been able to appreciate the unusual nature of his teaching, notably in the domain of natural philosophy. Today, we know of eleven disputations defended for the sake of exercise (*exercitii gratia*), which were associated with De Veno's lecture courses and therefore written by himself. Of these, nine treat topics in physics, one in metaphysics and one in politics. Furthermore, there is one set of theses *pro gradu*, for obtaining a master's degree, over which De Veno presided and whose contents he will presumably not have (entirely) written by himself.[75] Finally, we have a curious published list of "errors and contradictions" that De Veno claimed to have spotted in Justus Lipsius' best-selling political treatise, the *Politica* of 1589, whose Dutch translation was in 1590 published nowhere else than at Franeker. At the end of this *Syllabus of Errors and Contraditions*, which a handwritten note on the only extant copy from the New York Public Library dates to 1604, De Veno – who here identifies himself as "professor of politics" – offered to work his arguments into a full treatise if the Frisian authorities sponsored his proposals. Evidently, they did not, for nothing is known of such a book.[76]

When we put all these texts next to each other, it becomes obvious that De Veno was not at all the Platonist that earlier historians have believed him to be. Rather, his philosophy may be characterized as an attempt to combine an Italian approach to natural philosophy with the exigencies of Protestant theology and metaphysics. With this combination, he is both the direct precursor of, and the major influence on, his student Gorlaeus.

The importance attached to theology is clearly formulated in De Veno's premise that there can exist only one single truth, which has been revealed in the Sacred Scriptures. Given the uniqueness and unity of truth, it is illegitimate, so he argues, to claim that Aristotle was right philosophically but wrong theologically. By taking this view, De Veno followed in the footsteps of a number of contemporary German Protestant philosophers whom he frequently cites, such as Otto Casmann (1562-1607), Rudolf Goclenius (1547-1628) and Nicolaus Taurellus (1547-1606). These authors had recently staged a battle against the double-truth doctrine of colleagues they accused of following Averroes and according to whom certain statements could be philosophically true while being theologically false. In order to remove the tension between philosophy and theology, these German philosophers had tried, each in his own way, to align the two disciplines and had thoroughly reformulated metaphysics, logic and natural philosophy in the process.[77] That their reformed philosophies contained notions and principles that violated Aristotle's teaching seemed acceptable: they felt that, as a pagan author, Aristotle had clearly been ignorant of the truths of revelation. Quite evidently, De Veno attempted to insert his own teaching efforts into this larger enterprise. In fact, like the German authors he admired, he borrowed a whole series of non-Aristotelian doctrines from Italian natural philosophers, notably from Girolamo Cardano and Julius Caesar Scaliger, as well as from chemical authors of the Paracelsian tradition.

It is characteristic of this intellectual setting that the first disputation of De Veno's physics course opens with the issue of how to reconcile the conflicting authorities of Holy Scripture and philosophy. We are told that since the day in which Adam and Eve bit into the forbidden apple, our cognitive faculties have been hampered and our knowledge has been insecure – a conviction that, as we have seen, was shared by Gorlaeus.[78] Whoever wants to overcome these shortcomings is invited to turn to biblical revelation, to experience and observation, as well as to reason. In this enterprise, 'physics' (a term that, we recall, was at the time equivalent to 'natural philosophy'), is of great assistance. Although De Veno's definition of physics is aligned with contemporary textbooks ("physics is the contemplative science of natural bodies, insofar as they are natural"), the theological and medical uses to which he directs this discipline make it assume new and often decidedly anti-Aristotelian overtones.[79] As for theology, De Veno insists that "the sacrosanct word of God" must constitute the textual starting-point for the natural philosopher, as "Aristotle's physics is imperfect."[80] The fact that De Veno lists Adam, Noah, Solomon and other Old Testament figures among the "authors of physics" reveals that he, like other Protestant authors, believed in the existence of a "Mosaic physics" – a physics, in other words, that was contained, in part implicitly, in the account offered in Genesis of the creation of the universe. In this respect, it is revealing that he refers to the prolegomenon of Otto Casmann's recent *Cosmopoeia Christiana* (1598), where it is explained why "Aristotle must cede to Moses."[81]

In the second disputation, "On the principles and causes of natural things," De Veno defines three constitutive principles of things natural. These are not the Aristotelian principles of matter, form and privation, as could have been expected from an Aristotelian natural philosopher, but instead matter, form and spirit. Following Cardano, De Veno defines spirit, which thus replaces privation, as the efficient cause that brings about the merger of matter and form into a substance, and which furthermore inheres in the latter.[82] Nor is matter pure potentiality, as De Veno's Aristotelian colleagues continued to maintain, because it possesses its own body, "albeit a most imperfect one." This bodily nature explains why matter does not desire a form ("for it desires nothing of that, which it has").[83]

The third disputation, which deals with the "first affections of body" – that is, with motion, rest and time – is no longer extant, but we do possess the fourth disputation *De infinito et loco*. It is a disputation in which the influence of Protestant theological needs on the development of natural philosophy is particularly evident. In this disputation, we encounter Otto Casmann once more, whom De Veno follows in denying that any physical object can be infinite in the sense of lacking either limits or a middle, as God is the only actual infinite. As far as place (*locus*) is concerned, only created beings (*entia*) have a place, whereas God, whose essence is infinite, cannot be placed.[84] At this point in his exposition, De Veno turns to a cluster of concepts that was of particular importance to the thorny issue of the Eucharist: can a body – such as the bread and wine of the Last Supper – transubstantiate into blood and wine, while retaining its old appearance and attributes, as the Catholics and Lutherans maintained (though offering different explanations for this process), or does the Eucharist not involve any such transformation, as Calvinists and Zwinglians protested (although once more for different reasons)?

De Veno also caters to Calvinist needs when he addresses the issue of the relation of body, place and quantity. He defines "place" according to the "most learned and subtle Scaliger" as the "space of the thing or body that is placed, and which is contained inside of the surrounding body."[85] Like Gorlaeus after him, De Veno thus accepts Scaliger's consequential rejection of Aristotle's concept of "place" (as a kind of skin that envelops the object) and accepts the alternative proposal of defining the place of a body as the quantity of general "space" that is occupied by that body.[86] Invoking the arguments of Casmann and of the famous Paduan philosopher Jacopo Zabarella (1533-1589), De Veno furthermore argues that the accident of "quantity" cannot be separated from the body itself. From this he concludes, like many Calvinists at the time, that the Catholic doctrine of transubstantiation cannot be true.[87]

With the fifth disputation, *De mundo in genere*, we leave the realm of the "affections" and turn to the physical bodies themselves. The disputation begins by defining the world (*mundus*) as a body that contains heaven and earth and all that is in them. There is no world soul, as the Platonists believe, as the world "is governed by God's most noble spirit."[88] Very much like other contemporary Protestant thinkers, and particularly Calvinists, De Veno attributes much that used to be relegated to secondary causes directly to God's agency. The disputation is, for the rest, rather inconspicuous. However retrograde it may seem to us, there is nothing unusual about either his explicit rejection of Copernicus' heliocentric model or the insistence, against Aristotle, that the world is not eternal, but was created 5561 years before the disputation took place, and hence in 3957 BC.[89]

Among the extant physics disputations, the eighth comes next, and it addresses the elements. Elements are defined, quite traditionally, as "corporeal essences, individuated according to species, subject to change, out of which all mixts are constituted and into which they are resolved."[90] This definition shows not the least trace of the atomism that would constitute the core of Gorlaeus' metaphysics and physics. Indeed, for De Veno, elements remain "the first bodies that can be generated and corrupted," just as they had been for Aristotle, which means that they can be transformed into one another and dissolve into higher forms.[91] A clear departure from Aristotle is, by contrast, constituted by the doctrine that there are not four, but only three – or possibly just two – elements.[92] On this issue, we find once more strong doctrinal overlaps with Gorlaeus. Like his student, De Veno excludes fire from the list of elements. In his eyes, it is a mere "meteoron" – a phenomenon occurring in the stratum of air.[93] The three remaining elements are defined by their respective degrees of warmth (warm, temperate, cold), which are their primary affections, and by three degrees of humidity (wet, humid, dry), which are described as their secondary, passive activities.[94] These three elements are, however, not on a par, because unlike earth and water, air never enters into the composition of natural bodies but fills all empty spaces in the universe and functions as a carrier of heavenly heat.[95] This doctrine, as we have seen, returns in identical form in Gorlaeus. In all bodies, De Veno continues, it is the element of earth that provides the shape of the substance, contains the heavenly "signature" and nurtures the "seeds."[96]

Similar, though not identical, doctrines are broached in an unnumbered disputation "About air" (*De aëre*) of the same year. This time, air is clearly excluded from the list of elements, though it is defined as a "simple body." The reason De Veno offers for this elimination is that in the beginning, God created heaven and earth without needing air as an original ingredient.[97] After considering briefly the views of Scaliger, Goclenius, Taurellus, Justus Lipsius and Lambert Daneau (Danaeus) on the qualities of air, De Veno concludes that no substantial transmutation of air into either fire or earth is possible.[98] There can be no doubt that this set of theses, which De Veno himself calls "a disputation against the views of many Aristotelians," is directly inspired by the writings of Girolamo Cardano.[99] In his *De subtilitate* (1550), Cardano had developed a theory that had first been adumbrated in Aristotle's *Meteorology*, book IV, where it is proposed that natural substances are made up exclusively of earth (the principle of dryness) and water (the principle of wetness), which mixed under the influence of celestial heat. "All recognizable substances in our world contain these two elements," Aristotle says there, "and are to be assigned to one or the other according to the proportion in which they contain earth or water."[100]

From the late fifteenth century onward, the two-element theory had attracted the attention of Paduan physicians and philosophers, and commentaries on *Meteorology* IV had begun to proliferate. Girolamo Cardano, a Padua-trained philosopher-physician, developed the two-element theory into a veritable cosmology. Like De Veno after him, he defined elements as those bodies that could enter into mixtures, so as to form physical bodies. From the traditional list of the four elements, he excluded fire, which according to him was not a substance at all, and air, which he thought was clearly a substance, but not one that could mix with others. The function of air, he maintained, was that of carrying celestial heat down to the terrestrial realm.[101] Once again, as we have seen in our previous chapter, this is also Gorlaeus' view, and we are now in a position to identify its proximate as well as its remote source.[102]

That De Veno was acquainted with Cardano's physics is evident, for he mentions him with approval in another disputation in which the student is asked to defend the following thesis: "Is there any elementary fire existing underneath the lunar sphere? We deny it with Cardano."[103] In yet another disputation, De Veno also denied that book IV of Aristotle's *Meterology* was correctly named and argued that it dealt not with meteorological phenomena at all but with perfectly homogeneous mixtures.[104] In so doing, he sided with Alexander of Aphrodisias (2nd c. AD), who had stated that *Meteorology* IV was in truth a treatise on perfect mixtures, and with Italian authors such as Agostino Nifo (1473?-1538) and Pietro Pomponazzi (1462-1525), who named this Aristotelian work *liber de mixtis* and *liber de mixtione*, respectively.[105]

This two-element theory constitutes an important bridge to early modern atomism, because if earth is identified with the principle of dryness and water with the principle of wetness, it becomes much more difficult to maintain the traditional theory according to which all elements can be transformed into one another. We can see in Cardano and even more clearly in Scaliger how natural the next step was, namely of thinking of these two material principles in terms of unchanging particles and of imagining their union as the special coming together of minute pieces of matter.

The eminent historian of atomism Kurd Lasswitz ended his commentary on Gorlaeus' two-element theory with the words: "It would seem as if Gorlaeus had been the first who denied the transmutation of water into earth."[106] But as we have just seen, this doctrine had first been developed by Gorlaeus' teacher De Veno, who based himself on a north-Italian two-element theory, which in turn was due to a re-interpretation of Aristotle's *Meteorology* IV.

Given the intimate link between the redefinition of the elements and the theory of mixture, it is a fortunate coincidence that the penultimate extant disputation of De Veno's physics course treats "Of the generation and corruption of mixtures."[107] Our Franeker philosopher defines 'mixture' as "the mutation of the elements by the spirit for the sake of the production of a mixed body."[108] This definition, which had been adumbrated in the second disputation, is once again not Aristotle's, but Cardano's. The same is true of the view that this 'spirit' – which in the disputation *De mundo in genere* had been identified with "God's most noble spirit [...] governing" the world – is the efficient cause of mixtures, while the instrumental cause is "heavenly heat."[109] De Veno further defends the view that the quality of cold is never responsible for mixtures, but has limited agency inasmuch as it moderates heat through a reaction (*reactione*).[110] Here, he relies once more on Cardano's two-element theory, for he writes that the material of all mixtures is "the elements insofar as they are humid and dry," that is, consist of water and earth, "for these are the accidents that necessarily accompany matter."[111] Unlike his pupil Gorlaeus, who would defend the view that mixtures are merely accidental conglomerates of indivisible atomic units and thus *entia per accidens*, De Veno argues in a more traditional manner that in a mixture, new forms arise "out of the potency of matter."[112] Still, his position is not strictly Peripatetic, as he rejects both Aristotle's and Averroes' idea that the forms (that is, the specific qualities) of the elements are strengthened or weakened in mixtures, as "simply false."[113] What happens instead is that the "union of the primary qualities, being the product of their mutual action and reaction," produces a specific temperament (*temperamentum*) – an Aristotelico-Galenic term, as we have seen, which Gorlaeus would also employ.[114] As far as corruption is concerned, De Veno offers a technical explanation that is developed in response to the French physician Jean Fernel's (1497-1558) theory of putrefaction.[115] Natural corruption is the 'resolution' of the mixture into its original elements. It is caused by the influence of ambient heat, which increases the natural heat of the mixture, opens up its outer parts and thereby leads to the escape of the enclosed humidity. In the case of organic beings, this also leads to the loss of vital heat. What is left behind quickly grows cold and soft – and dies.[116]

The last extant disputation of the physics course treats of "the rational soul and its faculties." Traditionally, as we have seen in the context of Gorlaeus' two treatises, the various souls – the vegetative soul shared by all living beings, including plants; the sentient soul, shared by all animals; and the uniquely human rational soul – constituted the concluding and crowning topic of natural philosophy. De Veno states as a premise that on this question, all ancient philosophers had been mistaken.[117] He prefers to rely on Christian authors, notably on Thomas Aquinas, whom he quotes frequently, and on Thomists such as Crisostomo Javelli (c. 1470-1538), Thomas Bricot (d. 1516) and Arcangelo Mercenario (d. 1585). He is particularly interested in what might be called the soul's causal definition. As for the efficient cause, De Veno argues that all Greek philosophers had failed to understand that God was the immediate manufacturer of all souls. In the particular case of the human rational soul, it has neither a material nor a formal cause; because it is itself the substantial form that defines 'man' ("the form that informs 'man' is the substantial form of man"). As for its final cause, he explains that it lies in "all the operations of the soul."[118] Following in the footsteps of the Renaissance philosopher and logician Thomas Bricot, De Veno argues that the soul has no material or composite aspect, but is a formal being (*ens*) that is incorporeal and yet subsisting.

Unlike some contemporaries, he rejects the view that in humans, the three abovemention types of soul – vegetative, sentient and rational or intellective – exist as independent entities. Instead, he subscribes to what is called the 'unicist' account when he writes that there exists only one, rational, soul that possesses a threefold function.[119] This one and only human soul can be studied either on its own – as an immortal and self-sustaining immaterial entity – or in conjunction with the body, of which it is the "first act" and the "informing form"; only the second aspect belongs to natural philosophy, whereas the first is treated in metaphysics and theology.[120]

Of those disputations that are unrelated to De Veno's physics course, one consists of a set of seventeen "famous questions" that a candidate for the masters title in philosophy disputed under De Veno's presidency in 1605 and thus a year before Gorlaeus' enrollment at Franeker.[121] Although these questions and the answers given to them are few in number and extremely short, they provide a concise overview of De Veno's principal philosophical concerns. The candidate, who begins with ethics, first declares himself to be closer to Stoic and Platonist positions than to Aristotle's, not least because the former are more compatible with Holy Scripture.[122] Next, he turns to metaphysics, where he raises a crucial question that takes us to the core of the philosophy of his student Gorlaeus and to the point of the latter's disagreement with Stellingwerff: "Is the subject of metaphysics the intelligible, inasmuch as it is intelligible, or instead being [*ens*] insomuch as it is being [*ens*]?"

To this question, which in some way summarizes the two intellectual paths one could choose at Franeker, if one wished to cast a new basis for philosophy, his answer is as clear as it is important in our present context: "The first position has been defended by some neoterics, but we defend the latter thesis against them."[123] The 'neoterics' alluded to are obviously the Ramists, who were well represented at Franeker, notably in the person of De Veno's own teacher, Lollius Adama.[124] In defending an ontological definition of metaphysics, De Veno showed his preference for an approach to metaphysics that was at the time developing in Protestant Germany, and to which his own student Gorlaeus was to make a noteworthy and innovative contribution.[125]

De Veno's intellectual preference is evident also in the subsequent question, which attacks Heizo Buscher (1564-1598), a philosopher who belonged to the so-called Philippo-Ramist current. Against Buscher, De Veno's candidate affirms that no essential properties can be removed from a body without a concomitant loss of its essence. With this issue, he enters anew the thicket of the Eucharistic controversies, in which the possibility of separating a given substance – bread or Christ's body, respectively – from its properties was the central issue. In fact, De Veno's master candidate rebuts a range of Lutheran and Catholic authors. Among the latter, we find Cardinal Bellarmine, who is accused of having argued wrongly that a body could be in several places at once without filling space.[126] To encounter this name in a Franeker disputation carried out under De Veno is of course quite striking: nothing about the standard rebuttal of the Cardinal's much cited anti-Protestant work, the *Disputationes de controversiis Christianae fidei adversus hujus temporis haereticos* (which had come out in various instalments from 1581 onwards), could have made the audience suspect that De Veno was personally acquainted with this famous inquisitor, whom he had faced as the *consultor* of the inquisitorial court during his Roman trial.[127]

Moving on to physics, the candidate asserts that prime matter is an incorruptible body and, once again, that there exists no elementary fire under the moon.[128] Tycho Brahe – with whom we have already learned that De Veno's colleague, the mathematician Adriaan Metius, had worked in 1594 – is invoked against Aristotle's view that comets are phenomena generated from and in air.[129] In the remaining *quaestiones* of the disputation, finally, the candidate postulates that Aristotle had also been wrong about creation, about the highest good, about time and about the matter of the heavens, which is identical to the matter of the sublunary sphere.[130] This last thesis is of course revealing, as it documents the fact that De Veno anticipated Gorlaeus' much-debated rejection of the difference between sublunary and supralunary elements![131]

There are, finally, two extant disputations that are entirely unrelated to natural philosophy. The first deals with a subject belonging to public law. Given that in the Aristotelian university tradition, public law was understood as a political topic that belonged to the realm of practical philosophy, it was natural that De Veno would also have been expected to address it. The *Dissertatio politica de magistratu* of 1606 deals with the powers and functions of magistrates. It asks, among other things, about the personal qualities required of magistrates and their powers in the domains of war, politics and religion. De Veno relies heavily on Jean Bodin (1530-1596), the so-called father of state sovereignty. He defends a type of measured absolutism, stating that the prince stands above the people, but the law above the prince. However, the prince is not bound by any specific law (which he can change) but only by natural law. His powers are derived directly from God who is the *causa efficiens prima*, in contrast to the society of men, which represents only the *causa efficiens secunda* – an idea that we also find in Bodin.[132] Particularly important is De Veno's affirmation that the magistrates, not the religious authorities, should watch over religious practise and doctrine.[133] The crucial point is that with this view, as we have adumbrated above, De Veno would have fallen within the Arminian camp, which had only the year before began to insist – much to the displeasure of the orthodox anti-Arminian camp – that the punishment of sins, the settling of doctrinal disputes and the protection of a certain liberty of preaching and biblical interpretation was the privilege of the civil authorities, not of ecclesiastical ones.[134]

The last of the extant disputations defended under De Veno is entitled *De signo et signato*. Its topic, the relation between "sign and signified," is defined in the opening thesis as a subject matter that belongs exclusively to metaphysics, although many of the theses discuss questions that belong to logic. One of the key works plundered for arguments is in fact the *Problemata logica* of the Marburg philosopher Rudolph Goclenius (1547-1628). But since Peter Lombard's twelfth-century *Sentence Commentaries*, whose fourth book constitutes the *locus classicus* for this question, the relation between sign and signified had been also a theological issue. In De Veno's disputation too, these implications quickly come to the fore: "All the Lutherans err gravely when they claim that the sign is always at the same place as the signified."[135] The central issue at stake is, as in so many other disputations of this period, the interpretation of the Eucharist, or, more precisely, the presence of the body and blood of Christ in the consecrated bread and wine, which the Catholics and the Lutherans affirm, albeit for different reasons, and the Calvinist Protestants deny. For them, the real presence of Christ is not in the consecrated bread and wine, but occurs in the spirit of the believer during the act of consumption. To prove the local separation of sign and signified is thus a central concern for Calvinist theologians and philosophers alike. Typically, the defendant in the disputation insists that for a concept to capture the *ens* that is signified by it, it must be spatially separated from it. The ubiquitarians (who maintain that the risen Christ is ubiquitous in the same way as God the Father and can thus be equally present at all altars in the world simultaneously) therefore err in assuming that in the Eucharist there exists a double sign, namely the external sign of wine and bread, and the inner sign of the body and blood of Christ. De Veno's student is asked to insist, in this disputation, that the latter are merely the signified, but that the signified can never be internal to the sign.[136]

This disputation demonstrates exceptionally well how key theological concerns dictated the alignment of metaphysics, logic and physics with confessional doctrines. Unusual about this disputation is that its theses were not De Veno's – as would have been the case with all other disputations he chaired – but had been written by the well-known German philosopher Clemens Timpler, as the postscript declares.[137] In its dedication, the defendant, a certain Augustinus Arnoldi, identifies himself as a student from a prestigious institution we encountered before: the Steinfurt Academy (officially called the Gymnasium Illustre Arnoldinum), which lay in a town close to the Dutch border. Arnoldi mentions among his teachers not only the philosopher Timpler, but also the liberal but controversial theologian Conrad Vorstius (1569-1622), about whom we will hear more in the pages to come.[138]

This Steinfurt link is noteworthy for several reasons. Between its foundation in 1588 and the establishment of the University of Groningen (1614) and the Illustre School at Deventer (1630), both of which were nearby, Steinfurt's Gymnasium Illustre was one of the foremost institutions to provide the nascent Dutch Republic with Calvinist ministers. In those decades, many students from the eastern Dutch provinces went to Steinfurt to get at least a part of their education from its distinguished faculty. Otto Casmann, whose name we have already repeatedly encountered in De Veno's disputations, taught at Steinfurt between 1589 and 1595, and Clemens Timpler, his successor, lectured there from 1595 to 1624.[139] Their combination of a Ramist methodology with a reformed Aristotelian metaphysics and physics influenced the teaching at Franeker in numerous ways.[140] We have also already heard that the rector of Leeuwarden's Latin School since 1607, Edo Neuhusius, who was also Otto Casmann's nephew, was a Steinfurt alumnus.[141]

However, in the second half of the year 1610, this serene relation of mutual benefit turned sour. The reason for this sudden change was the nomination of Steinfurt's professor of theology, Conrad Vorstius (whom we have just encountered in the dedication of De Veno's student Arnoldi), as the successor of the recently deceased Jacob Arminius at Leiden University. This appointment exacerbated the battle between Remonstrants (Arminians) and contra-Remonstrants (anti-Arminians), which culminated in 1618-1619, when the Synod of Dort condemned the Remonstrants and banned Vorstius from Dutch soil. The episode and its eventual outcome, to which we will have to return below, left deep traces in the evolution of Dutch Calvinism and at the same time alienated Steinfurt and the Dutch academic establishment from one another.

We heard earlier that the Arminian issue erupted violently at Franeker in the year of De Veno's rectorate, in 1609, and have suggested that his involvement in it may have been the cause of his temporal removal from both the rectorate and his professorial chair. Seen in this perspective, one is left wondering whether the appearance, in 1604, of a Steinfurt student of Timpler and Vorstius who disputed under De Veno on matters carrying heavy theological implications points to a more profound institutional and an intellectual bond between De Veno and the Steinfurt academics. Given the scarcity of the printed material of that period, we cannot decide this issue. However, we have seen that De Veno cites Otto Casmann and Clemens Timpler frequently and with approval in his disputations. We have also seen that De Veno seemed to share Vorstius' political preference for a government that kept the churches as well as dogmatic disputes under their control.

In sum, then, what must we think of Gorlaeus' teacher Henricus de Veno? Irrespective of the secretive aspects of his biography, it appears from the extant disputations that De Veno's teaching presented a noteworthy combination of recently developed philosophical positions. We have seen that, contrary to some of his colleagues at Franeker, he was not interested in Ramism but was instead attracted by the theologically motivated ontological concerns of such German philosophers as Goclenius, Taurellus and later Casmann. Furthermore, he was the only Dutch professor whose teaching reflected the cosmology and theory of matter of Girolamo Cardano and, to a lesser extent, of Julius Caesar Scaliger. Thanks to his colleague Adriaan Metius, he was furthermore aware of Tycho Brahe's observations of comets and used them to deny the immutability of the celestial spheres and the existence of a non-elemental ether. While it would, a few decades later, no longer be uncommon for teachers of natural philosophy to mention the novel results of the empirical disciplines, De Veno seems to me to have been the only Dutch philosopher to have done so in the opening years of the seventeenth century.

3.6. GORLAEUS' DEBT TO DE VENO, CARDANO AND SCALIGER

There is undeniably a fresh air of modernity about De Veno's disputations, whose printed versions date to the years 1603 to 1606, the latter being also the year in which Gorlaeus enrolled at Franeker. They combine an 'Italian' approach to natural philosophy with the most up-to-date Protestant doctrines on metaphysics and physics. True to his motto, "I have no authorities" (*authoritates non habeo*), De Veno followed, besides Aristotle, a variety of theological, Platonist, Stoic, medical and naturalist authors, using as his main criterion the agreement of their respective views with Holy Scripture, reason and experience (in this precise order).[142]

This link between theology, metaphysics and physics is comparable to what we have found in Gorlaeus' writings, the main difference being that De Veno's *disputationes* allow us to recognize more directly the theological motifs and sources behind his philosophical choices, as well as the authors and books that nurtured them. By contrast, Gorlaeus almost never mentions his sources (a fact which explains how he could have been mistaken for an experimental scientist).

As for the theological concerns, they are particularly evident in the *Disputatio metaphysica* defended in 1604 under De Veno by Augustinus Arnoldi, the abovementioned Steinfurt student whom Clemens Timpler had sent over to Franeker.[143] We recall that this "metaphysical disputation" addresses the issue of the physics of the Eucharist, which was one of the key levers by which confessional reasons brought about doctrinal adjustments in metaphysics and physics at Protestant universities.[144] How such an adjustment worked can be observed in the disputation *De infinito et loco*, in which De Veno's preference for a general space (*spatium*) over a localized place (*locus*) is overtly linked to the Calvinist interpretation of the Eucharist and hence to the need to rebut both the Catholic doctrine of transubstantiation and the Lutheran doctrine of ubiquity. More precisely, quoting a number of Protestant philosophers, including Goclenius, Taurellus and Casmann, De Veno tries to demonstrate that it is impossible for a body to be separated from its place (*locus*), because the *locus*, being a quantity, is necessarily tied to the body.[145] In this instance, as in others, De Veno (like Gorlaeus after him) tried to propagate an ontology that substituted accidents by essential attributes. Moreover, if all things exist in a specific place (because place, *qua* quantity, is tied to them), it follows that prime matter must be considered "a substance, a body, and incorruptible."[146] What is adumbrated here was to become much more explicit in Gorlaeus: there is no space for potentialities in this world; whatever is, is fully and actually so, and matter is represented by the atoms that make it up, which are incorruptible units.

Importantly, this tight link between physics, metaphysics and theology explains why De Veno and his student Gorlaeus could share the conviction that the study of natural philosophy was capable of improving the situation of our fallen souls: "thanks to the knowledge of physics, we may arrive at the knowledge of God and his power."[147] Importantly, the physics that De Veno has in mind is not Aristotle's, which is "not worthy of a Christian, nor of a philosopher," but one that is in accordance with the "sacrosanct word of God," on the one hand, and with observation, on the other.[148] It is the "sacrosanct word," for example, that demonstrates why Aristotle's axiom that *ex nihilo nihil fit* is wrong.[149] Here, we have one more of the sources of Gorlaeus' drawn-out attack on this axiom.[150]

But De Veno introduced Gorlaeus not only to the world of Protestant metaphysics and to ontology in particular, but also to a view of natural philosophy that was directly inspired by Cardano's *De subtiliate* (1550) and somewhat less by Scaliger's *Exercitationes exotericae de subtilitate* (1557). By the early seventeenth-century, both works were well known and repeatedly reprinted north of the Alps; and yet, when we compare De Veno's views on matter theory with those of his Dutch contemporaries, we find that he is the only philosopher who dismisses Aristotle's physics in favor of Cardano's.[151]

In chapter 2, we analyzed Gorlaeus' two-element theory. In the present chapter, we have seen that he received it from De Veno, who in turn had taken it from Cardano. The latter's *De subtilitate* develops an idea that had first been adumbrated in the fourth book of Aristotle's *Meteorology* and had, from the late fifteenth century onwards, attracted the attention of Paduan physicians and philosophers. The Italian philosopher Francesco Piccolomini summarized this view as follows: "Cardano affirms in his Book on mixture that mixts do not consist of fire, nor of water, but exclusively of earth, water and heavenly heat."[152] This is essentially the elemental theory that Gorlaeus was to adopt and merge with his own atomism. Whether he needed to read Cardano or learned these positions directly from De Veno, who repeatedly quoted Cardano with approval, cannot be decided on the basis of the texts we have. Either way, De Veno's disputations help us understand the appeal that this theory seems to have had for the two men: the definition of mixture as the "mutation of the elements by the spirit for the sake of the production of a mixed body" allowed one, in their eyes, to posit "God's most noble spirit" as the efficient cause of mixtures and "heavenly heat" as its instrumental cause.[153]

Despite his acceptance of Cardano's and De Veno's theory of elements and mixtures, Gorlaeus modified it considerably by joining it to his own atomist ontology. Neither Cardano nor De Veno had postulated the existence of atoms. The latter maintained that elements are "the first generable and corruptible bodies," attacked those who believed that the "forms of the elements remain intact in the mixt" and taught that the *temperamentum* was merely the accident of a substantial *forma mixti*.[154] Although Gorlaeus, too, believed in the existence of such a common 'temperament', in his eyes, the elements *did* remain intact in the mixts, namely as atoms; nor was the temperament an accident of a new substantial form:

> Bodies that are mixed out of earth and water have no essence that is distinct from either earth or water. For they are something composite. But we have said before that no composite is anything else besides its parts or has another essence than these parts.[155]

However, let us recall that Gorlaeus' theory of mixture also assumed that the atoms participating in a mixture exchanged their non-essential qualities (notably hot and cold) and thereby brought about the above-mentioned common 'temperament'. Importantly, this precise theory has parallels with Scaliger's *Exercitationes exotericae de subtiliate* that are too close to be accidental. Although Gorlaeus inherited Cardano's matter theory from his teacher De Veno, it would appear that he personally inclined much more to Cardano's nemesis Scaliger, who is in fact the only modern philosopher he mentions in his work. In fact, he heaps praise on him, calling him "the most subtle of all philosophers." It is also Scaliger whom he quotes in his boisterous entry in the *Album amicorum* of his friend Engelbert Egidius van Engelen, as has been mentioned earlier. In fact, he may have chosen both the title *Exercitationes philosophicae* and its division into 'exercises' in honor of Scaliger's much cited book.

Admittedly, De Veno had also occasionally relied on Scaliger, following the latter, as we have seen, in rejecting the Aristotelian definition of 'place' (*locus*) in favor of a general 'space' (*spatium*). But Gorlaeus went further by also responding positively to Scaliger's corpuscular interpretation of the *minima naturalia*, with which the chameleonic Italian polemicist had explained a whole range of natural phenomena. We recall that Scaliger chose to ignore the categorical difference between a genuine Aristotelian *minimum* (which denotes the lower quantitative limit of matter capable of sustaining a given substantial form or of performing an action) and a corpuscle (which describes an autonomous piece of substantial matter). This allowed him to claim, among other things, that the *minima naturalia* of an anvil are so densely arranged that they cannot be further compacted even by a hammer; that fire is stronger or weaker depending on whether its particles are closer or farther apart; that the density of the *minimae partes* generally explains the specific properties of substances; that fire divides earth into its *minima naturalia*; and that some substances have round or oblong *corpuscula*.[156] In fact, among the sixteenth-century authors who pretended to remain faithful to the spirit of Aristotle's philosophy while transforming the notion of *minima* in the direction of independent material units, or atoms, Scaliger is probably both the most extreme and the most prominent.[157]

But for all his apparent admiration for Scaliger, by replacing the latter's ill-defined *minima* with fully fledged atoms, Gorlaeus took a radical further step. We recall from above that he accepted Scaliger's famous definition of mixture as "the motion of the minimal bodies towards mutual contact, so that a union is brought about," adding that "by minimal bodies, we mean individual atoms."[158] In 1629 the influential Wittenberg professor of medicine and chemist Daniel Sennert was to subject Scaliger's definition to a similar transformation, by claiming that what Scaliger had "without doubt" intended by his 'minima' were Democritean atoms.[159] However, both Sennert and Gorlaeus were mistaken in their interpretation; but while Sennert realized full well that he was, but decided to invoke Scaliger simply to buttress his case, Gorlaeus may possibly have believed that his own reinterpretation remained faithful to the underlying idea formulated by that much-admired author.

If Scaliger represents the acme of the sixteenth-century north-Italian tendency to understand Aristotelian *minima naturalia* as independent corpuscles, Gorlaeus represents the moment in which minimism officially converted to atomism and in so doing became an overtly *anti*-Aristotelian doctrine.[160] It is this anti-scholastic turn away from Aristotle that utterly distinguishes Gorlaeus from Scaliger.

At this point in our investigation, the following question arises: from whom did Gorlaeus take his own overt atomism? Certainly not from Scaliger, who despite his own corpuscular tendencies – or, if you like, his latent crypto-atomism – repeatedly condemns this doctrine, insisting that

> if the forms remained intact, mixture would in truth be a mere heap. [...] These would be true Democritean atoms. Hence they would be certain quantities, not parts of a single mixt, but each would be a totality to itself.[161]

But Gorlaeus claims, as we have seen, precisely the contrary: each atom, being an *ens per se*, is "a totality to itself."[162]

While our analysis of the teaching in natural philosophy that Gorlaeus enjoyed at Franeker, enriched by references to Cardano's and Scaliger's matter theories, has helped us find the origin of Gorlaeus' theory of elements and even of some of his specific corpuscular explanations, the provenance of his atomist ontology, this capping stone of his physics, has so far eluded our grasp. So as to find this last element, we must continue to follow him through his life and move on to Leiden.

3.7. GORLÆUS AT LEIDEN

Gorlaeus' Franeker education and the authors to whom he was introduced there explain a great number of elements of his own metaphysics and natural philosophy. But they fail to answer the following questions: What were the reasons that persuaded Gorlaeus to develop De Veno's matter theory and Scaliger's minimism further into an explicitly atomist doctrine? And why did he make physical atomism depend on a fully developed "prima philosophia de ente," an ontology that included God, humans and physical bodies alike? Finally, why did he, at such an early stage in his academic education, complete two treatises the publication of which would have befitted a university professor, but not a beginning theology student?

In order to explain these questions, we must try to understand what happened to Gorlaeus as he moved on to Leiden. Unfortunately for us, there exists a temporal gap between his graduation from Franeker, probably in 1609 or 1610, and the moment, on 23 April 1611, when "David Gorlaeus, from Utrecht, aged 20, student of theology, [lodging] with Magdalena, daughter of Laurentius," enrolled at Leiden's theological faculty.[163] This date of April 1611, precisely a year before he passed away, is relatively late. What had Gorlaeus done in the meantime?

Before other documents surface, this question must unfortunately remain unresolved. All we currently have of this period is Gorlaeus' entry into the *Album amicorum* of Engelbert Egidius (which will be discussed below), dated 25 June 1610.[164] Jaeger has concluded from this entry that Gorlaeus lived at Leiden the year before he started studying.[165] But in contrast to just about all other inscriptions, Gorlaeus does not indicate a place name, and there is no indication that Engelbert Egidius (who had just returned from his academic pilgrimage and started working as a minister in Arnhem) had in 1610 spent time in Leiden. Nor would it have been financially attractive to move to a university town without registering as a student. In fact, Gerben Wierda has postulated that it was more likely that the two men met in Franeker, and as evidence takes the entry in Engelbert Egidius'

Album written by the reverend Christoph Hardenberg at Franeker, in September 1611.[166] As Hardenberg's inscription starts in Hebrew, Wierda proposes, it might have been that Engelbert Egidius was privately studying Hebrew with the Franeker professor Johannes Drusius.[167] Alas, this interesting hypothesis can currently not be confirmed. By the time Hardenberg signed Egidius' *Album*, Gorlaeus was already studying theology in Leiden. Furthermore, Egidius' name does not feature in Franeker's student register; it would have been highly unusual and financially unrewarding for a student to spend so much time at a university without formal enrollment – the time lag between Gorlaeus' and Hardenberg's entries being fourteen months. However, as Wierda has also documented, Arnhem's church council was looking for a new minister in early 1610 and had cast its eye on a certain Henricus Meiling, who was at the time minister in a village near Leeuwarden. It may conceivably have been the case that Egidius traveled to negotiate the position and that he may have combined such a trip with a visit to Franeker to discuss with Drusius his commissioned work on a new translation of the Book of Hebrews.[168]

Whatever happened in 1610, and irrespective of the date of Gorlaeus' move from Friesland to Leiden, what seems clear is that he had started working on his own philosophy back home. His friend Stellingwerff's reference to the ambitious works that he was engaged in seems to confirm this. It would also explain why the manuscript of the *Idea physicae*, although apparently completed, was not published, and why Gorlaeus chose to absorb its physical doctrines into his almost megalomaniacal metaphysical treatise. The new influences that he worked into the *Exercitationes*, and which are absent from the *Idea*, would suggest a change of mind. This change of mind may, as we shall now see, have been caused by Engelbert Egidius, in whose *Album* Gorlaeus immortalized himself. It is certainly also due to theological and political circumstances, which we shall now have to examine.

The likely causes behind his change of mind carry two names that are of great importance to the early history of the Netherlands: the Arminian Crisis and the Vorstius Affair. There exists an extremely precious testimonial which, though written down thirty years after the composition of Gorlaeus' two works, is of extraordinary precision. The context of this testimonial takes us to the Utrecht Crisis, that violent controversy over Descartes' teaching that started in 1641 and led to the condemnation of Cartesianism at that university. As has been adumbrated before and shall be told in some detail in chapter 4, that quarrel was initially triggered when the professor of medicine, Henricus Regius, had one of his students defend a thesis according to which "man was a composite being" (*ens per accidens*). When he was attacked for this view as well as for his Cartesian positions more generally by Utrecht's leading theologian, Gijsbert Voetius, Regius defended himself by arguing that this was neither his own thesis, nor Descartes', but had been lifted out of Gorlaeus' *Exercitationes*. This explanation, instead of soothing Voetius' anger, only served to enrage him further. In a tremendously long-winded sentence, which, however, overbrims with extremely useful factual information, the theologian penned the following denunciation:

> The paradoxical claim [about man being an *ens per accidens*] made [...] by Taurellus (who was called an atheist physician by the Heidelberg theologians in their judgement on Vorstius' *De Deo*, which they sent to the delegates of the Synod of Holland in 1610), and which, due to the imprudence of youth, our compatriot David Gorlaeus took up in his *Exercitationes philosophicae*, a book he wrote in a moment when, beginning his theological studies or rather preparing himself for them, he was attacked by doubts and hesitations [...] is contrary to truth.[169]

These densely argued lines contain several precious pieces of information, which we now have to disentangle and interpret, as they contain an abundance of clues concerning the background and thrust of Gorlaeus' *Exercitationes*.

To begin with, Voetius states that Gorlaeus wrote his *Exercitationes* not during his "theological studies" but "rather" while "preparing himself" for them. This confirms us in our idea that Gorlaeus must have been working on his treatise before enrolling at Leiden's Faculty of Theology in April 1611. But whether this 'preparation' took place in Leiden or back in Friesland, remains unclear. Furthermore, in the testimonial just quoted, Voetius mentions an affair surrounding a work called *De Deo* by Conrad Vorstius, a figure who has already repeatedly been mentioned. He was the Steinfurt theologian whom Leiden University had chosen as Arminius' successor in 1610, but who was forced to leave Leiden almost immediately after his arrival in 1611 because the general uproar caused by his theological positions had become unmanageable. As we shall see in detail below, the affair came to a head in the very period that Gorlaeus was in Leiden. The name 'Taurellus', in turn, refers to the German philosopher Nicolaus Taurellus, whose theological ontology – first presented in his *Philosophiae triumphus* (Basel, 1573) – did indeed exert a certain influence on Vorstius and was in some quarters viewed as a dangerously heterodox philosopher.

Voetius' claim that there is a direct link connecting Taurellus, Vorstius and Gorlaeus is not only suggestive but, as we shall see below, convincing. Besides Voetius' assertion of 1641 and the evidence that a comparison of Gorlaeus' doctrines with the writings of Taurellus and Vorstius can yield, we fortunately also possess some direct biographical evidence to corroborate this affiliation. There are two persons with whom we know Gorlaeus to have been acquainted in the period 1610-1611. Both strengthen the hypothesis that he was intellectually affiliated with the Arminian camp and that the intellectual debt that Voetius suggested in 1641 corresponds to the truth.

The first figure is Engelbert Egidius van Engelen (c. 1584-1642), whose *Album amicorum* is today kept at Leiden's University Library. This *Album* contains an exuberantly juvenile entry by Gorlaeus, dated 25 June 1610. Engelbert Egidius had started this *Album* in 1606 after finishing his theological education at Leiden and before setting off for his *peregrinatio academica* through France, Switzerland and Germany. Dozens of (partly very famous) academics inscribed themselves between February 1606 (Leiden, the point of departure) and April 1609 (Heidelberg, the last stop of his academic tour). The *Album* contains a single inscription from 1610 (by Gorlaeus – but where did the two meet?) and one from 1611 (by Hardenberg in Franeker).[170]

But let us turn to Gorlaeus' entry, which constitutes right now the only autograph we have of his hand. The album page is reproduced in Figure 15. The boisterous entry states, in translation, as follows:

> The home (*patria*) of a prudent man is wherever he is at ease (Julius Scaliger, in his Poems).
> This is the sum of all my sums: It is stupid, lazy and of leaden madness to omit action and grow old with words.
> I have written this entry with my own hand as a perpetual memory of myself and as a token of my most friendly disposition towards Engelbert of Engelen, of Arnhem in Gelderland, much commended by his true virtue, solid erudition and moral integrity, much-praised candidate in true philosophy and singular friend of mine, on 25 June 1610.
> David Gorlaeus of Utrecht.
> Motto: Virtue shall find its way.
> Even if everyone holds the opposite view, truth must be defended. And one must judge one's own teacher in the same way as one's greatest enemy. This be your eternal rule!
> Live, and remember us.[171]

The tone of the entry suggests that Gorlaeus and Engelbert Egidius knew each other well, but as mentioned, we know nothing about the origin of their relationship. Egidius was eight years older than Gorlaeus. Born in Arnhem, he had enrolled at Leiden for literary studies in 1601 and had later moved on to theology. In 1605, he defended a set of theological theses under Franciscus Gomarus. In 1609, he became minister in Oosterbeek, a village in the neighborhood of his hometown, and subsidiary preacher in Arnhem. He lived and preached in Arnhem, although he never received a regular appointment there. The ministers and the church council being torn between the conflicting currents within the Reformed Church, all preachers in Arnhem had to sign, in 1614, an Act in which they had to

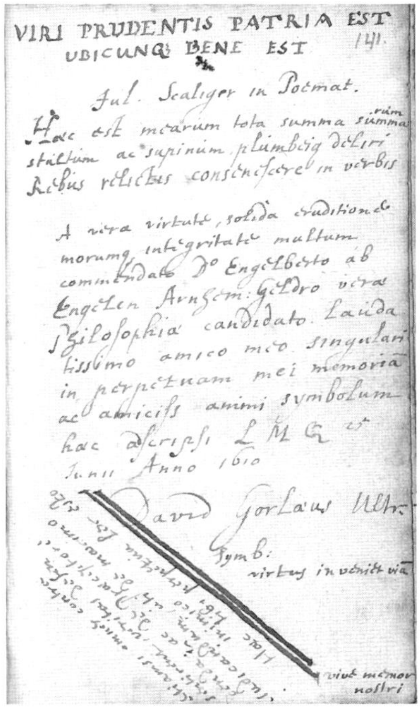

FIG. 15: The only known autograph by David Gorlaeus is his inscription in the *Album amicorum* of the theologian Engelbert Egidius van Engelen, dated 25 June 1610. This album reads like a "Who-is-who in early seventeenth-century Protestant theology." Gorlaeus is one of the youngest persons to have an entry in it. (University Library Leiden, MS Papenbroeck N. 21, fol. 141r. Courtesy of Universiteitsbibliotheek Leiden)

promise, among several other things, "to avoid the particular names of Arminians, Remonstrants and Contraremonstrants from the pulpit."[172] Although apparently cautious and non-polemical, Engelbert Egidius was according to church historians "in heart and soul a Remonstrant";[173] and so, when the provincial synod of Gelderland, held in Nijmegen in 1619, required of all ministers to subscribe to the anti-Remonstrant articles of the Synod of Dort, Engelbert Egidius was found to hesitate just a bit too long, and was consequently sacked. For the rest of his life, he was counted among the Remonstrants. In fact, he was reinstated later in his life as an official preacher in Arnhem's Remonstrant church.[174]

Despite the paucity of the available information, Egidius provides us with a highly suggestive clue concerning the stipulated intellectual between Taurellus, Vorstius and Gorlaeus. Taurellus' ground-breaking *Philosophiae triumphus*, from which Gorlaeus was to take the idea of man as an *ens per accidens*, was republished in 1617 in a very unlikely place, namely Arnhem, by the town's only printer, Jan Janssen (Janssonius). This second edition, which is identical to the first down to the details of the layout, and up to its inclusion of Taurellus' original dedication letter to a nobleman, does not so much as hint at the reasons that led to its production. Interestingly, however, it can be demonstrated that in the years that he was preaching in Arnhem, Engelbert Egidius was collaborating with Janssen's press.[175] In 1615, for example, Janssen published the Dutch translation of a French treatise narrating the conversion, the year before, of a French Capuchin to the Protestant faith. This work had been translated by Engelbert Egidius, who dedicated the book to Arnhem's governors.[176] It is therefore plausible to assume that it was he who cajoled Janssen into republishing Taurellus' voluminous philosophical work, all the more as this philosophically and theologically charged work was so clearly associated with the Calvinist current with which Egidius was associated.

To be sure, Taurellus had already featured in De Veno's disputations of 1603-1606, though not prominently. But given Taurellus' much stronger influence on Gorlaeus (which will be documented below), Gorlaeus' friendship with Engelbert Egidius and the latter's presumable involvement in the publication of the second edition of Taurellus' *Philosophiae triumphus*, is not at all implausible to see in Gorlaeus' inscription in Engelbert Egidius' *Album* the reflection of an important intellectual bond, in which generally Arminian inclinations were transformed into a more explicit sympathy for the embattled theologian Vorstius and for the alleged source of some of his metaphysical ideas, namely Taurellus.

Gorlaeus' second acquaintance from the period 1610-1611 is even more adventurous. We recall that when he enrolled at Leiden University, he indicated as his address the house of "Magdalena, daughter of Laurentius." Six weeks later, on 4 June, a certain "Rudolphus ab Echten" signed up at the same university indicating the same address. What no historian seems to have realized is that the young and affluent nobleman Rudolph van Echten (1592-1643) was the same figure who had a few months before contributed to that immense Socinian scandal at Franeker mentioned above.[177] Van Echten, who was half a year younger than Gorlaeus, had first been sent to the Steinfurt Academy by his parents. In May 1610, he enrolled at Franeker, the registers of which record, as the university's 1188th student, one "Rudophus ab Echten, nobleman from Drenthen, coming from the Steinfurt school."[178] In the same year, a number of other students reached Franeker from Steinfurt who had all, in one way or another, imbibed Socinian sympathies already before arriving and probably also bristled at the theologically intolerant atmosphere at the Frisian institution and at the dogged and inquisitorial behavior of Sijbrand Lubbert, the professor of theology.[179]

Lubbert, who was at the time working on a book against Socinianism – the heterodox, anti-Trinitarian sect set up by Fausto Sozzini, which was at the time perceived by all established confessions as the most dangerous theological movement on the European scene – seems to have peppered his lectures with invectives against Sozzini. For reasons that are not easy to comprehend given the foolishness of the enterprise and the dangers involved, a number of students, mostly from

Steinfurt, decided in late 1610 or early 1611 to publish Sozzini's *De officio hominis Christiani* (which has been mentioned earlier). Although the author of the treatise was not mentioned explicitly and the place of publication was given as the "City of Peace" (*Irenopolis*), it transpired soon enough that this work (which in hallucinatory madness pretended to enjoy the approval and privileges of both the Pope and the Spanish king!) had in truth been published at Franeker. This discovery caused an immense scandal with the local authorities; for, in this book, Sozzini reiterated just about every heresy for which he was so notorious: he reproached the Protestants for not having carried the Reformation to its obvious conclusion by abolishing such Catholic additions to the original faith like the Trinity, the divinity of Christ and the role of the Holy Spirit and by adhering to erroneous views about baptism, the Eucharist, predestination and the absence of free will.[180]

Once rumors had sprung up that this heretical tract had been published in Friesland, church, secular and university authorities intervened speedily. In May or June 1611, student chambers were searched; the printer was unmasked; and upon finding epistolary evidence and having cross-examined students, it was established that a group of students, who had for some time taken an active interest in Sozzini's theology and had even established epistolary links with Socinians elsewhere, were behind this *editio princeps* of Sozzini's *De officio*. Not all, but most of these students had a Steinfurt connection and, to make matters more explosive, they had a connection to Vorstius, who was at the time awaiting official approval to succeed Arminius in the Leiden chair of theology. Of these students, Bernard Fockenbeck had been at Steinfurt before enrolling at Franeker and in 1609, when Steinfurt was not yet a suspicious place, he had (like Arnoldi in the De Veno disputation, discussed above) dedicated a Franeker disputation to his Steinfurt teachers Vorstius and Timpler. Heinrich Welsing's connection was even stronger: he had been Vorstius' amanuensis for no less than five years, before moving over to Franeker in September 1610.[181] Jacob Omphalius, in turn, whom interrogations unmasked as the driving force behind the Socinian publication, had since his arrival from Steinfurt repeatedly clashed with Lubbert and Franeker's Academic Senate.[182] Certain documents suggest that Omphalius was also acting as a tutor to the somewhat younger baronet Van Echten, whom he somehow got involved in the plot to publish Sozzini's *De officio*.

As the authorities cracked down on the group, its members quickly dispersed. While the Germans fled back to Steinfurt, the Dutch went to Leiden, which, as one Frisian observer noted, threatened to become "a hospice and asylum of such people."[183] Aemilius Trebatius, the chief Frisian member of the Socinian group and probably the author of the explanatory postscript to *De officio*, had removed himself from Frisian jurisdiction to Leiden in June, although he only enrolled at Leiden the year thereafter.[184] The same holds true for the young Van Echten, who also rushed to Leiden where, as we have seen, he inscribed himself in philosophy and arts on 4 June. Keeping a low profile in the hope of going unrecognized, he changed some details about his identity: remaining silent about his nobility, he downplayed his age (indicating 18 years instead of 20) and altered the name of his birth-place: the university official transliterated his mumblings as "Drechtanus," a funny mix-up of the place name "Echtanus" ("from Echten"), the homonymous estate of which he had been proprietor since 1607, and "Drentanus" ("from Drenthe"), the name of his province of origin.

Like Van Echten, Gorlaeus was – at least on his mother's side – of noble birth; and as the records of the 1606 Franeker book auction indicate, his purchasing power was noteworthy. One must therefore assume that the lodgings that the two rebellious upper-class students shared in the house of "Magdalena, daughter of Laurentius," belonged to the upmarket type. However, much more interesting about this roommate is the light that it sheds on Gorlaeus. Had Gorlaeus, too, been involved in the plot to publish Sozzini? Had he, too, left for Leiden as a consequence of the discovery that Irenopolis was in reality Franeker?

Unfortunately, we do not know. At any rate, it is evident that Gorlaeus knew, frequented and – judging by his writings – sympathized with figures who in more or less radical ways sided with the Arminian faction, supported Arminius' supposed successor, Vorstius, and read unsavory philosophical authors such as Taurellus or such dangerous heretics such as Sozzini. When Voetius, in the convoluted historical declaration, cited above, writes that Gorlaeus had developed his philosophical views in a period of juvenile disorientation, it must now be obvious what he meant by this: Gorlaeus had come under bad influences and was siding with theologically aberrant groups.[185]

Having briefly introduced the theme of Arminianism, Conrad Vorstius and Nicolaus Taurellus from the perspective of Gorlaeus' *mauvaises fréquentations* at Franeker and Leiden, let us now approach them also from the point of view of the history of philosophy and theology and insert Gorlaeus' own thought into that intellectual landscape.

3.8. JACOB ARMINIUS AND THE BEGINNING OF THE ARMINIAN CONTROVERSY

The Arminian controversy and the violent tones it quickly assumed are nowadays hard to understand and to explain. However, in the years 1610 and 1611, in which the bulk of Gorlaeus' manuscripts was written, it would have been impossible for any intellectually active Dutchman to ignore it; and if one was, like Gorlaeus, about to become a theologian, and specifically at the University of Leiden where the controversy had broken out and where it continued to have its epicenter, it would have been impossible not to side with one of the two quarrelling factions. Moreover, in 1611, only 21 new students enrolled in theology at Leiden, and all of them were expected to choose their side in this issue.[186]

Let us therefore briefly explain the origin of the controversy and the point it reached in Gorlaeus' days.

The Reformation had been spreading in the Low Countries slowly since the 1520s and more rapidly after 1550, with the propertied middle classes in Flanders and Brabant being drawn to the new faith more quickly than the peasant populations, partly as a reaction to the absolutist tendencies of the Spanish Habsburgs that threatened their political and economic liberties. As the state fought the Reformation as a heresy, it seemed quite natural for the discontented groups to oppose the church together with the state. Prince William of Orange, the king's lieutenant (*stadtholder*) in Holland and Zeeland, began in 1564 to intercede with King Philip II of Spain in favor of a more flexible approach towards non-Catholics, but to no avail. An explosive combination of Spanish inquisitorial intolerance, economic hardship brought about by excessive additional taxation and religious fervour led, in 1566, to local rebellions that degenerated into iconoclastic attacks on churches and monasteries. In reaction to the fast-spreading "Protestant fury," the Spanish crown dispatched an army led by the notoriously inflexible Duke of Alba, who quickly cracked down on the local nobility and citizenry. The result of the ensuing hardening of the situation is well known: it is the Dutch rebellion against the Habsburg Empire, which became irreversible once Philip II had declared William of Orange an outlaw in 1580, and the States-General responded to the king with their Act of Abjuration.

It would take until 1648, and the length of the so-called Eighty Years' War, for this break to yield a fully sovereign Dutch state. What matters in our present context is that in the process, it was not the Lutheran type of Reformation that gained the upper hand in the Low Countries, but its Calvinist variant. In the early period of the rebellion, various cities in Flanders emulated Geneva, by setting up what is often described as "Calvinist Republics." When Alessandro Farnese, Duke of Parma, conquered Antwerp and other towns in Flanders and Brabant, 'heretics' were given two years to leave the territory. About 200,000 persons are estimated to have followed this unkind invitation, migrating north and thereby strengthening the Calvinist component of what was slowly solidifying into the Netherlands.[187] Although 'Calvinist' is not the label that the Dutch Protestants used for themselves, it has become the historian's name for the more severe and doctrinally rigid form of Protestantism that came to dominate the Netherlands, given that it drew theological inspiration from Geneva's reformer and was institutionally linked to

Geneva and Heidelberg. Its theology entailed a tightly structured church government with regional units (classes) and provincial and national organs (synods) and with a high degree of mutual supervision and doctrinal homogeneity.[188] The pillars on which this homogeneity was to be built were the Heidelberg Catechism (1563) and the Belgic Confession (1561).

The controversy to which we must now turn and which was to tear the Dutch Protestant Church apart in the early seventeenth century is best approached from these two documents. It all started off with that thorny issue of predestination – the question of whether God, in his omnipotence, eternity, omniscience and immutability, had from all eternity decreed all events that took place in the world; and, more narrowly, whether our individual election to paradise or condemnation was likewise predetermined.[189] The Heidelberg Catechism did not address the issue directly, but the *locus classicus* is question 54 and its answer:

> *What do you believe concerning the holy and catholic church of Christ?* I believe that the Son of God from the beginning to the end of the world, gathers, defends, and preserves to himself by his Spirit and word, out of the whole human race, a church chosen to everlasting life, agreeing in true faith, and that I am and forever shall remain a living member thereof.[190]

In turn, article 16 of the *Confessio Belgica* stated:

> We believe that when Adam's entire offspring worked its own destruction through the sin of the first man, God showed himself for what He is, namely merciful and just. Merciful, since He rescues and delivers from doom those whom He in his eternal and unchangeable counsel out of sheer grace has chosen in Jesus Christ, our Lord, regardless of their works. Just, since He leaves the others in the fall and doom, which they have brought down upon themselves.[191]

Before it reached the Netherlands, the debate concerning predestination has had a long theological prehistory (going back to the Apostle Paul), and a much shorter Calvinist trajectory, which started with Geneva's reformer Jean Calvin and his successor, Theodor Beza. Neither need detain us here. Suffice it to say that the theological debate that sprung up in the Netherlands in the first years of the seventeenth century quickly grew in intensity and bitterness and eventually led to a schism within the Dutch Protestant Church. The Dutch debate as such and the liberal, latitudinarian interpretation of predestination, however, took their name from the theologian Jacob Arminius (1559/60-1609; see Figure 16).

Born as Jacob Hermansz. in the small Dutch town of Oudewater, Arminius obtained his training first at Marburg (1574-75) and then at the University of Leiden

(1576-1582), which had been founded the year before his arrival.[192] In fact, Arminius was but the thirteenth student to enroll at this new institution.[193] A grant from Amsterdam's merchant guild allowed him to continue his theological studies in Geneva (1582-1587) under Jean Calvin's stern successor, Theodor Beza (1519-1605). He interrupted his Genevan studies for a longer stint in Basel (1582-1584). With a letter of recommendation by Beza, addressed to the Amsterdam city counselor and praising his intellectual talents, Arminius returned to Amsterdam. However, for his return, he took a rather unusual detour, visiting Italy for several months. His opponents would later allege that he had gone to Rome to kiss the pope's slippers. Whether he had seen the pope from far away or up close is uncertain.[194] What is certain, by contrast, is that he stopped in Padua, where he attended a number of lectures by the famous logician and natural philosopher Jacopo Zabarella. Upon his eventual return to the Netherlands, he was ordained a minister in Amsterdam in 1588 and served as a pastor until his appointment as university professor in 1603.

Even in the years of his ministry, however, Arminius aroused the suspicion of some of his colleagues for his alleged support for a number of non-orthodox positions, also with respect to the bothersome question of predestination. For the present purposes, it is this specific problem that interests us most. "Arminius never

FIG. 16: Jacob Arminius, engraving by Willem van Swanenburg, from *Icones ad vivum delineatae et expressae* (Leiden, 1609). This oldest dated portrait is analyzed in Tolsma, "*Iconographia Arminiana*," 241.

rejected predestination," Marius van Leeuwen has protested, "but probably, from the beginning of his ministry, inclined to an infralapsarian idea of it."[195] But what exactly does this 'inclination' imply? The term 'infralapsarianism' refers to the Fall (*lapsus*) of man that was caused by Adam and Eve's disobedience to God and their subsequent expulsion from Paradise. Ever since, and by virtue of the transmission of this original sin from generation to generation, humankind has existed in a fallen state. There was, of course, that central figure of the redeeming incarnation of God, Jesus Christ, who with his death on the cross had atoned for mankind's sinfulness. However, he had not washed away all sin *tout court*. After all, it was known through scriptural revelation that only a few were elected to be forgiven and obtain eternal life in Heaven, or Paradise, while the majority of humans would be left in their fallen state and end up in Hell. So much, in very rough outline, was commonly agreed upon between the various Christian confessions.

What was disputed, by contrast, was whether the individual human being could contribute anything to his or her personal salvation, and if so, how. The Catholic Church emphasized faith, to be sure, but also stressed the importance of good works – not just prayer and moral conduct, but also acts of benevolence, including quite notoriously the payment of money to the Church itself in the purchase of indulgences. The Calvinist standpoint, by contrast, insisted that salvation could be obtained *sola fide*, 'through faith alone'. Now, given that faith was not something that in the eyes of Calvinist theologians could be acquired through an act of will, which was not free anyway, but was instead a gift of God, it seemed to follow that the individual couldn't do anything about his or her election, which was entirely in the hands of God.

This doctrine of predestination – one's election (or otherwise) even before one had the chance of proving oneself worthy of it – did not only rely on theological considerations, but rested above all on philosophical logic. A God that was defined to be both eternal and omniscient could not but predict *ab aeterno* who would be saved and who would not. His decision (which was referred to as *decretum* or 'decree') as to whom he would save from damnation and whom he would leave in a fallen state preceded the birth of the individual in question. We have seen, above, the precise way in which the Belgic Confession couched this understanding.

Within the dire logic of predestination, there were additional theological distinctions that could be battled out. We have just mentioned that Arminius was suspected from early on to be favorable to the 'infralapsarian' interpretation of the doctrine of predestination over the more orthodox position of 'supralapsarianism'. This distinction refers to the moment at which God took his 'decrees' regarding individual election or otherwise: had it been before the Fall of man ('supralapsarian') or afterwards ('infralapsarian')? This question, over which much ink was spilled, had theological implications regarding God's relation to the Fall itself (had he foreseen or possibly even willed it?), but, as is obvious, also strictly logical implications. If God is defined as one and inseparable, and if he lives and thinks in eternity and not in actual, creaturely time, it is difficult to explain a change of mind in response to temporal events taking place under an apple tree.[196]

This brief survey of the two views on predestination allows us to return to Arminius. When in 1602, two of the three theology professors at Leiden were swept away by the plague, Arminius' name came up as a possible successor of Franciscus Junius, one of the plague victims. Notably, Johannes Wtenbogaert, Arminius' friend since their common Genevan days, who had in the meantime become court chaplain to Prince Maurice, lobbied for Arminius. Franciscus Gomarus, Leiden's only surviving theologian and thus *professor primarius*, nurtured severe doubts concerning Arminius' suitability, accusing him of Pelagianism, which in this case came down to the charge of ascribing free will to man. However, after a conversation with the proposed candidate, he agreed to supervise Arminius' doctoral degree, which he granted in July 1603. With this degree in hand, Arminius was permitted to take the Leiden chair. Less than a year later, however, the theological strife that would soon give birth to the Arminian current within Calvinism was already in full swing between the two theologians. In two sets of student disputations, both defended in 1604, their differing views on predestination became manifest.[197]

It was really the definition and, as it were, the mechanism of predestination that divided the two men. Once again, their differences of opinion are so subtle, and from a modern theological viewpoint so minimal, that it is difficult to understand the agitation that they managed to stir up and which would soon turn into veritable hatred. While attempting to avoid the heresy of Pelagianism, which deemed man capable of obtaining salvation by his own nature, Arminius at the same time also tried to avoid the blasphemous implication that God was the originator of sin – an implication that seemed to follow from the notion that God had from the beginning, and possibly even before the act of creation, decided whom to save from an original sin that was yet to take place and whom to condemn. In a rather complex formulation, he therefore defined predestination as "the decree of God's good pleasure in Christ, by which he resolved within himself from eternity, to justify, adopt, and bestow with eternal life believers, whom he decreed to bestow with faith, to the praise of his glorious grace."[198] However roundabout and diplomatic this formula may seem to us, Arminius' insistence that "this is the will of God that everyone who sees the Son [Christ] and believes in Him will have eternal life" attributed, in the eyes of Gomarus and the divines that followed him, too much of a role to faith in Christ as a precondition for election, which therefore seemed to become in some sense conditional. To be sure, Arminius did not make salvation depend on the combination of faith and good works, as Catholics did, but he did attribute to the individual the capacity to resist salvation, by deliberately counteracting the faith that God had bestowed on him. For this reason, his theological opponents soon convinced themselves that "by diminishing the role of God and pleading for human freedom, Arminius distanced himself from the Reformed confession" and notably the *Confessio Belgica* and the Heidelberg Catechism.[199]

Soon, the theological debate assumed political overtones, for, as it heated up and threatened to get out of hand, Arminius proposed that it should be solved, or at least supervised, by the civil authorities, whose task he felt it should also be to make sure that a certain amount of dissent – a certain 'latitude', as the English tongue would soon put it – could be tolerated. *In necessariis unitas, in non necessariis libertas, in omnibus caritas* ("Unity in things necessary; liberty in things non-necessary; and in everything charity") was in due time to become a famous slogan among Arminians. Arminius' so-called Erastian approach (named after the sixteenth-century theologian Thomas Erastus, who had pleaded that civil authorities, not the Church, should punish the sins of the faithful), was indeed to become a trademark of the Arminian faction. Important sympathizers such as the theologian Wtenbogaert and the lawyer Hugo de Groot developed it further.[200]

However, precisely this apparently tolerant, state-supervised approach to religious discussion was seen by Arminius' opponents to be yet another sign of his non-committal leniency towards Catholic, Jesuitical and Spanish interests and thus as treacherous and damaging gestures of prostration to the enemy of the young Dutch Republic.[201] This accusation of Romewardness and treason was of course grave and turned Arminius into an object of incessant insults, as pasquinades were placarded all over the city of Leiden, attacking or ridiculing him. And quickly, as the controversy grew bigger, it spilled beyond Leiden's boundaries, soon involving preachers, universities and magistrates all over the country and also abroad. After all, Gomarus himself had claimed that the issues involved were worth a civil war: "province against province, church against church, city against city, citizen against citizen!"[202] (See Figure 17)

From a theological point of view, Arminius' *Declaration of his Sentiments with Respect to the Predestination* of October 1608, delivered to the States of Holland and West-Friesland, is generally considered the clearest and also boldest statement of his views.[203] In it, Arminius insisted that God was not in the first instance an immutable judge, electing or damning *ab initio*. His "first precise and absolute decree" had not been the predestination of individuals to salvation or damnation, but "the salvation of sinful Man."[204] He had prepared "in a sufficient and efficacious manner the means which were necessary for repentance and faith" – which left some space for individuals to repent and believe.[205] Finally, God had not decided beforehand who would repent and believe, and therefore be saved, although he "knew from eternity" who would do so. In other words, Arminius divided God's foreknowledge from his will, thereby leaving space for individuals

Fig. 17: This political cartoon of 1618, called "The Arminian Dung-Cart," shows the "so-called Reformed," pulled by two "discordant" horses, on the "way to Rome" (as the inscription between the hooves of the first horse indicates). Johannes Wtenbogaert is shown on the box seat. Jacob Arminius, Petrus Bertius, Conrad Vorstius, a certain Anabaptist, and other known non-orthodox reformers are depicted behind him. Johan van Oldenbarnevelt shows the way with his torch (left margin); two Jesuits watch the scene with approval (foreground center). (Private collection)

to grasp their chance and contribute to their salvation – or to lose God's grace in case of an obstinate refusal to accept it.[206] In a witty pro-Arminian pamphlet, *The Predestined Thief*, the criminal protagonist shrewdly argues along orthodox lines to the effect that he had no reason to mend or repent his ways, since he knew that his election was independent of his behavior, backing his arguments with well-chosen quotations from Calvin, Beza, Gomarus and Piscator.[207]

Arminius' opponent Gomarus delivered a speech in December of the same year, also before the States of Holland and West-Friesland, explaining why he deemed Arminius' views to be "unbiblical, heretical and confused."[208] The State subsequently urged the two men to come to an agreement during a reconciliatory meeting early the following year. But Arminius, who was already too ill from tuberculosis, had to return from the meeting to Leiden, where he passed away on 19 October 1609. Petrus Bertius (1565-1629), regent of Leiden's Theological College ("Statencollege") and an old friend of Arminius (as well as an acquaintance of Gorlaeus' uncle, Abraham), held the much publicized funerary oration, and began organizing the succession.[209] As for the oration, it depicted Arminius as a peaceful man whose life had been soured by envious enemies. Rather than putting an end to the whole affair, Bertius' oration only poured oil on the flames, provoking angry reactions first of all on the part of Gomarus himself, who speedily published his *Considerations Concerning Bertius' Funerary Oration.*[210]

Instead of quenching the debate, Arminius' death only marked the transition from an individual stand-off to a collective one. On 14 January 1610, the draft of a so-called *Remonstrance*, which was probably formulated by Johannes Wtenbogaert, was signed by 44 ministers and in June of the same year submitted in a somewhat altered form to the States of Holland and West-Friesland. What had begun as the view of single theologian, Arminius, had with this act grown into 'Arminianism' or, indeed, 'Remonstrantism'. Attempts by the Grand-Pensionary Johan van Oldenbarnevelt and the famous lawyer Hugo Grotius to bring about a climate of tolerance failed, or were rather seen as pro-Arminian latitudinarianism *avant la lettre*.

In 1617, Prince Maurice took the consequential step of openly siding with the anti-Arminians, when instead of attending the Sunday service of his court chaplain, the Arminian Wtenbogaert, he went to The Hague's Kloosterkerk, which had recently been occupied by the anti-Arminians. Soon, a veritable persecution set in. Johan van Oldenbarnevelt and Hugo Grotius were arrested. The former, condemned for high treason, was beheaded; the latter received a life sentence (but famously managed to escape from Castle Loevestein hidden in a book chest). A national synod was finally organized. However, the Synod of Dort (1618-19), though organized according to Arminius' Erastian ideas by the civil authorities, represented no attempt at reconciliation, but instead a severe crack-down on the Arminian faction, which was condemned as heretical (see Figure 18). Ministers who would not recant were expelled from the ministry.

The Synod of Dort was not the end of Arminianism, however. In the autumn of 1619, a number of exiled Arminians set up a Remonstrant fraternity in Antwerp, once more under the leadership of Johan Wtenbogaert. In 1621, Simon Episcopius wrote a confessional creed in order to give theological coherence to the exiled group and to rebut a series of attacks by opponents. The situation of the émigré sect only improved with the death of Prince Maurice and the succession of his brother, Prince Frederik Hendrik, as stadtholder in 1625. Wtenbogaert had been Frederik Hendrik's tutor and had maintained good contacts with him. As a consequence, the Arminians were allowed to trickle back into the Netherlands. In 1630, they opened a new church in Amsterdam and in 1634 even a seminary in which they could educate their own clergy. This allowed the Remonstrant camp to develop into one of the "three systematic models arising out of Protestantism, the Reformed, the Lutheran, and the Arminian."[211] Its rationalism, on the one hand, and tolerance, on the other, were to prove their effectiveness in later decades of the seventeenth century; not only in the Netherlands, but also in England, the United States and elsewhere.[212]

FIG. 18: Opening of the National Synod of Dort on 13 November 1618. The Remonstrants sit at the table in the middle. The president, Joannes Bogerman, is seated at the main table, in front of the open fire. The representatives of the Palatinate, Hessia, Switzerland, Nassau, Geneva, Bremen and Emden are seated in the benches on the right-hand side. The benches for the French remained empty. This engraving is based on a print by François Schillemans. (Private collection)

3.9. THE VORSTIUS AFFAIR

The Synod of Dort also sealed the fate of Conrad Vorstius, the theologian who had been appointed to fill Arminius' vacant chair at Leiden University. As we recall from above, the Utrecht theologian Gijsbert Voetius would later associate Gorlaeus' *Exercitationes* with one of Vorstius' works, which is why we must now take a closer look at this historical character.

For this, we must return to 1610. We have just heard that Arminius' death, while obviously putting an end to his personal struggle with his Leiden colleague and nemesis Gomarus, had also given birth to a wider movement. We recall that in the early months of 1610, 44 ministers had signed the so-called 'Remonstrance', which reiterated the main Arminian positions on predestination. Its so-called 'five points' asserted, in a succinct modern phrasing, that (1) election and condemnation on the Day of Judgement is conditioned by man's rational faith or non-faith; (2) the atonement, while qualitatively adequate for every man, is efficacious only for the man of faith; (3) without the assistance of the Holy Spirit, no person is able to respond to God's will; (4) grace is not irresistible; and that (5) believers are able to resist sin but are not beyond the possibility of falling from grace.[213]

At the same time, efforts were undertaken to appoint a successor to Arminius himself. Petrus Bertius, who had stirred up new emotions through his funerary oration, and Johannes Wtenbogaert, Arminius' old ally and preacher to Prince Maurice, pressed for the appointment of Conrad Vorstius (1569-1622).

Born in Cologne and trained in theology first at Herborn and then at Heidelberg, where he took his doctorate under Johannes Piscator, Vorstius (see Figure 19) had passed through Geneva, like all credible Calvinist theologians, where he had defended a number of theses under Beza. We have already encountered him in his later position, as professor of theology at the Gymnasium Illustre at Steinfurt, which – as we may recall – had strong connections with the Netherlands in general and with Franeker in particular.

It was not easy to cajole Vorstius into accepting Arminius' chair at Leiden. Aware of the factional fights at Leiden, he hesitated to place himself in such a vipers' nest. He had already in 1599 encountered problems with the self-appointed watchdogs of orthodoxy, who had accused him of Arianism, of Unitarianism and, what was the worst accusation possible, of Socinianism.

Since these terms will recur in the few pages to come, it is useful to get them out of the way. 'Arianism' refers to the belief, first propagated in the fourth century AD by Arius, that Christ is not truly divine and was created *ex nihilo* by God in a specific historical moment. This view obviously implied that Christ could not be considered on a par with God as a full member of the Holy Trinity. Further, an Arian would claim that only God is self-existent and immutable, while Christ is mutable. 'Unitarianism' can be related logically to 'Arianism': it is, roughly speaking, a system of Christian thought that derives its name from the doctrine that God, the Father, is a single personality, a doctrine that stands in opposition to the Trinitarians' view that the divinity is three-fold. It is customary to regard Michael Servetus (burnt in Geneva in 1553) as well as Lelio and Fausto Sozzini (respectively 1525-1562 and 1539-1604) as the fathers of early modern Unitarianism. Fausto Sozzini in fact gave his name to various Unitarian movements. As a label, 'Socinianism' emerged towards the end of the sixteenth century and was generally perceived as the new and most dreadful threat to Protestant unity. Sozzini's most important work, *On the Saving Work of Christ*, argued that Christ's death at the cross was in fact an exemplary case of (human) atonement.

The degree to which Vorstius must be considered a 'Socinian', a 'crypto-Socinian' or neither of the two need fortunately not be defined here. Whatever the truth of the matter may be, it is noteworthy that the label stuck. A full hundred years after the event, Gottfried Wilhelm Leibniz, in the introduction to his *Theodicy* of 1710, still associated Vorstius with the Socinians:

FIG. 19: A mid-seventeenth-century portrait of Conrad Vorstius, by an anonymous engraver. (Private collection)

> Even though there were no co-operation by God in evil actions, one could not help finding difficulty in the fact that he foresees them and that, being able to prevent them through his omnipotence, he yet permits them. This is why some philosophers and even some theologians have rather chosen to deny to God any knowledge of the detail of things and, above all, of future events, than to admit what they believed repellent to his goodness. The Socinians and Conrad Vorstius lean towards that side.[214]

The link between Vorstius and the Socinians has at times been seen to reside not only in doctrinal overlaps (such as a reduction of the role and divine status of Jesus Christ or a limitation of God's absolute powers, mentioned by Leibniz), but also in the historico-critical method of Bible exegesis that Vorstius first, and subsequently many Arminians, are said to have learned from Sozzini.[215] The key publication in that respect was Sozzini's *De auctoritate Sacrae Scripturae*, a book that he had composed in the late 1560s and first published under the name of a Jesuit, Dominicus Lopez. Because of its application of Lorenzo Valla's critical method and Matthias Flacius Illyricus' comparative method, *De auctoritate Sacrae Scripturae* has been described as the first modern case of a philological approach to the Bible.[216] Astonishingly enough, Conrad Vorstius, in the middle of all the allegations that he was a heterodox philo-Socinian, re-edited this very book in 1611 at Steinfurt, just before setting off to take the Leiden chair in theology that had been offered to him. In his defense, he stressed that this book had not aroused anyone's suspicions, not even when it was published at Basel in 1592.[217]

Vorstius' overall publication policy of 1610 and 1611 appears indeed completely counter-intuitive. Back in 1599, it had required protracted negotiations, a flood of writings and a severe interrogation at Heidelberg to clear Vorstius' name of the accusations of heresy raised against him. Having been through such turmoil before, he was understandably reluctant to change the relative tranquility of Steinfurt for the doctrinal turbulences at Leiden. That he eventually succumbed to the temptation to accept the Leiden chair turned out to be a fatal mistake; but his editorship of a Socinian work was outright incomprehensible. After all, his leanings towards the Arminian faction were outspoken, both politically and theologically. In an astonishingly frank and imprudent piece of political advice, which he offered in the dedicatory letter prefaced to his *Anti-Bellarminus* of early 1610, Vorstius told the Dutch States-General that they would be well advised to keep the Church under their full control and to make sure that theologians did not miss the true meaning of Christianity, which lay in living in faith and piety, and not in academic disputations and pamphlet wars. Such counsel, Vorstius continued, was necessary given that the evangelical churches had already suffered enough self-destructive fragmentation through the work of demagogical theologians. To avoid any further fragmentation, the state ought to allow for, and watch over, three types of freedom: freedom of conscience (*libertas conscientiae*), a native liberty (*nativa libertas*) in doctrinal interpretation and a freedom of preaching these interpretations publicly (*prophetandi libertas*).[218]

It deserves to be mentioned in this context that Vorstius' role in the evolution of the concept of a *libertas philosophandi* has been sorely neglected, although he quite clearly anticipated distinctions that are usually attributed to later Arminians like Philipp van Limborch. Let us point out that in his letter of 13 October 1611 to Isaac Casaubon, Vorstius also anticipates the important distinction between essential and non-essential doctrines when contesting Casaubon's demand for synodal restrictions on theological views. Vorstius demanded that only doctrines directly grounded in Scripture should be imposed, whereas freedom of interpretation must be guaranteed for all other doctrines.[219]

But to return to his dedication letter prefaced to his *Anti-Bellarminus* of 1610: his advice to the Dutch States-General added new arguments to the Erastian views of the Remonstrant faction. While it must have angered the anti-Remonstrants, it seems to have endeared him to the civil authorities, who were already negotiating with him about the Leiden chair. However, later in the same year, 1610, Vorstius also published an extended edition of a series of commented theological disputations previously defended at Steinfurt. It carried the title "Theological treatise on God, or: About God's nature and attributes" (*Tractatus theologicus de Deo, sive de natura et attributis Dei*). The cry of his first biographer – *ô librum natum in turbas!* ("ah, what a book born into trouble!") – captures but faintly the uproar this book created within the ranks of the anti-Arminians, who found its views befitting an "Atheist, pagan, Jew, Turk, heretic, schismatic, and ignoramus," but certainly not a Calvinist theologian.[220] Moreover, his opponents claimed with increasing insistence that Vorstius sympathized with, or even entertained direct links with, the Socinians. It seemed clear that he had taken a more than fleeting interest in Sozzini's views not only back in 1599, when he had been admonished, but again, or still, in 1610. Even to the theologically uninstructed reader, it was obvious that his writings displayed doctrinal overlaps with the Unitarian views of the Socinians, notably concerning the alleged difference between God's eternity and divinity and Christ's temporal and at best semi-divine status.

In the light of all these allegations, the campaign to block Vorstius' installation as professor gained increasing momentum and in the end became literally majestic as it involved a personal refutation of the doctrines contained in his *De Deo* by King James I of England and the public burning of his works in Oxford, Cambridge and London.[221] The English crown became very much involved in the affair with its ambassador to the Netherlands, Ralph Winwood, going about agitating against the '*monsterdier*', the "monstrous animal," as which Vorstius had been unmasked.[222] In his *Church History* of 1655, the English clergyman and historian Thomas Fuller narrates the prehistory of the English involvement in this affair. This is the character sketch he provides of Vorstius:

> this wretch did seek to stoop God to man, by debasing his purity, assigning him a material body; confining his immensity, as not being everywhere; shaking his immutability, as if his will were subject to change; darkening his omnisciency, as uncertain in future contingents: with many more monstrous opinions, fitter to be remanded to hell, than committed to writing.[223]

A pamphlet war was well underway in 1610, long before Vorstius actually moved to Leiden, and it reached new levels of intensity as he was finally appointed on 23 August 1611, a few months after David Gorlaeus had formally enrolled in the theological faculty.[224] Already by then, Vorstius' precariously embattled situation had become untenable: the discovery, discussed above, that some of his students (including Gorlaeus' roommate Van Echten) had published Sozzini's *De officio Christiani hominis* at Franeker and that through these students, Vorstius appeared to be in contact with Polish Socinians, undermined his credibility even in the eyes of most of his erstwhile supporters. What the latter had looked for was a liberal, but respectable theologian; what they got was instead a radical thinker who went far beyond anything that Arminius or his friends would ever have dared to think, let alone put on paper, and who moreover did not seem to be able to keep his own actions or those of his Steinfurt students under control.

When Vorstius' protestation that there was no such Socinian connection began to look implausible even to his Dutch friends, Leiden University's Academic Senate, which in electing him had braved immense political and ecclesiastical pressure, eventually yielded. With polemics threatening to become uncontrollable, its members decided to suspend Vorstius from his obligations, although they did not divest him of his appointment. Some sources declare that this suspension took place after merely three months of lecturing (which Gorlaeus would have been able to follow); other sources maintain that Vorstius was never even given the possibility to deliver a single lecture.[225]

From the city of Gouda, to which he had withdrawn in May 1612, he continued to defend himself in writing against numerous epistolary attacks. This uneasy state of affairs went on until the Synod of Dort, which cracked down on the entire Remonstrant current and specifically decided to ban Vorstius from Dutch territory. In the English translation of 1619, the final decree states as follows:

> this *venerable Synode* [...] hath with ioynt suffrages declared [...], that the said *Conrade Vorstius* (besides that concerning the fiue controuerted Articles, he defendeth and maintaineth the errors of the *Remonstrants*, rejected by this Synode) doeth in his latter writings but especialy in his Tractate, entitled, *Of God and his attributes*, make bold with, not one or two Articles of the reformed Religion, but most of the fundamentall heads of Diuinitie; namely, such as concerne the Trinitie of persons in the Godhead, the Simplicity, Infinitenesse, Immensitie, Essentiall Omnipresence, Omniscience, Omnipotency, Wisedome, and Immutability of the Essence of God; as also concerning the Creation, the Prouidence of God, the Hypostaticall Union of two natures in Christ, the full and perfect Satisfaction performed by Christ for our sinnes, the Iustification of man before God by Faith, and many other Articles particularised, as well by the most high and mighty King of *Great Britaine*, as by diuers professed Diuines. [...] So that it manifestly appeareth, that his intent was cunningly to make way for the secret instilling of the impious heresies of *Socinius* and others: and that he, under pretence of inquiring, doeth bestirre himselfe to seduce others.[226]

It is interesting to find among the signatures confirming the validity of the decrees of the Synod those of the violently anti-Arminian and anti-Socinian Sijbrand Lubbert, theologian at Franeker University, and of Gijsbert Voetius, Gorlaeus' former fellow-student, who was at the time pastor in Heusden and later professor of theology at Utrecht and a fierce anti-Cartesian polemicist who would for the rest of his life cite Vorstius' work as one of the worst theological offenses ever committed.

Whether Vorstius had been a sly and ambitious arch-heretic or rather a peaceful theologian who unwittingly became the victim of a situation of theological mass hysteria is hard to decide. While his opponents viewed his Socinian sympathies and his political advice to the Dutch political establishment as highly dangerous, the historian of Arminianianism, Harrison, writes: "It is impossible not to sympathize with Vorstius. He was not one of the square-browed, pugnacious theologians, who were so plentifully produced at that time and seemed born for the prize-ring."[227] At any rate, Vorstius left the Netherlands a broken man. Once again, there exist two versions of what happened to him. According to one, he first went into hiding and eventually moved to Friedrichstadt, where he died shortly after his arrival in 1622. According to the other, he became a teacher in Tönningen, and died there.[228]

But let us finally return again from Arminius and Vorstius to the subject of our book. It is evident that Gorlaeus witnessed the beginning of the Arminian controversy at Franeker, whose faculty had split already during the early phase of the standoff between Arminius and Gomarus, with its professor of theology, Lubbert, styling himself as the leader of the Frisian anti-Arminians. By enrolling at Leiden's theological faculty, however, Gorlaeus had now moved to the epicenter of this massive doctrinal earthquake. It goes without saying that it would have been impossible for a theology student not to take sides in the issue. In October 1610, for example, an overwhelming majority of 55 theology students out of 68 signed a submission to the Curators of the University and the States Provincial of Holland and Friesland against Vorstius' appointment.[229]

It is obvious that no student of theology could have remained agnostic in this matter. And all the evidence we currently have – doctrinal and biographical alike – indicate that Gorlaeus had chosen for the minority current, the Arminians.

3.10. THE LINK BETWEEN VORSTIUS' *De Deo* AND GORLAEUS' *Exercitationes*

It has already repeatedly been mentioned that thirty years later, in 1641, Gijsbert Voetius located the *Exercitationes* within this precise context, linking its doctrines to Vorstius' *De Deo* and to Taurellus' ideas. In this section, we aim to verify whether this allegation is true. Irrespective of our verdict on this issue, however, it should be clear that Gorlaeus wrote his two works in the years in which the Vorstius affair peaked; and that he did so at the two university towns at which the Remonstrant and Socinian debates were raging most violently. Whether this atmosphere influenced, or determined, or possibly even motivated Gorlaeus' work is what we must now examine.

In order to understand the link, suggested in Voetius' 1641 testimony, between Gorlaeus' ontological atomism and Vorstius' *De Deo*, we must first understand what it was that scandalized the anti-Arminians about the work of the German theologian. One of the main two points they attacked was Vorstius' attempt to find common ground with non-Calvinist Christian sects. His doctrinal flexibility must by the way not necessarily be read as evidence of a deeply felt religious tolerance on his part; it may have been motivated by the observation that the Protestant world was in a state of increasing fragmentation and by an urgent feeling that this process needed to be stopped in the face of mounting counter-reformational pressure. Vorstius' position on this issue could hardly have been expressed more clearly and forcefully than in his call for a State-enforced *prophetandi libertas*; i.e., a freedom of theological interpretation.[230] In his Introduction to *De Deo*, he moreover expressed the hope that his method and its results would lead to a moderation of the various stand-points and hence to a unification of Protestant churches.[231] The thrust of this argument is comparable to Taurellus' hope that a commonly acceptable metaphysical basis would allow for the development of a greater doctrinal consensus between the competing theological groups – and in turn to Gorlaeus' attempt to develop an ontology that could serve also as the basis of a *theosophia*. All of this is directly related to the second general trait of Vorstius' thought that aroused scandal among his doctrinal foes. Many theologians were taken aback by his use of metaphysics and physics, not only as a means of resolving theological issues, but as an outright instrument of salvation – in the eyes of the predestinarians not only an erroneous but also a clearly futile enterprise. This (meta-)physical approach to theology is probably where Taurellus' influence on Vorstius is at its clearest; it also sheds light on, and possibly even explains, Gorlaeus' attempt to enter the debate through the presentation of a worked-out ontology.

Vorstius' censured doctrines fall roughly into two groups.[232] The first, which will not concern us here, has to do with the (lesser) status of Christ vis-à-vis God; they are the ones that led to the charge of Socinianism and Unitarianism. The second group is related to the problem of predestination. Vorstius' double premise is that God's vindictive or punitive justice is not part of his essence, and that it is wrong to define faith as a form of confidence in the forgiveness of sins, for this is not its essence. The implication of this double premise is that the relation between God and humans is much more open than the anti-Arminians permitted; for if it is neither part of God's essence to punish, nor part of ours to be saved or condemned, it follows that we, in the temporal course of our lives, may be able to contribute to the sentence pronounced on us on the Day of Atonement.

How such a personal effort can be rendered compatible with divine omniscience is of course what needs to be explained. Vorstius' method for doing so relies on an ontological *Wesensbestimmung* of God, as the title of his embattled book indicates.

In essence, his *Tractatus de Deo* tries to define God's attributes and to deduce from them that God, despite his eternity, is capable of undergoing changes in time, and notably of changing his mind. Importantly, in changing his mind even God can be led by hopes and fears. Such actual changes of mind are mentioned in the Bible, whenever we read of God passing new laws. The most notable such case is the New Testament, which demands greater perfection of the faithful than the Old Testament.[233]

The upshot of this temporalization of God's actions is that it allows for an interaction between him and humans, and hence for a weakening of the dire logic of predestination. Vorstius' conclusions are in this respect compatible with Jacob Arminius' position: God's decisions, for Vorstius, are merely accidents of his being; his will is not unchanging, because it concerns events that are in time; God is spatially separated from what happens, because he is not everywhere present in being, but only through his actions; and for him, eternity is not an indivisible entity, but a mere succession of past, present and future. In fact, God does not know future events in the same way in which he knows past events; he can think one thing after another in the usual manner of a deliberation. For this very reason, it is logically possible that his decrees regarding matters that depend on his free will were not taken at the beginning of all eternity.[234]

Agreeable though many of these views will have been to the more daring exponents of the Remonstrant movement, in the eyes of his opponents, Vorstius committed the crime of physicalizing God, whom he treated like any other *ens*. The English censors, for example, listed as the first of Vorstius' heresies the following: "God is not essentially immense, nor simply infinite; but he is a quantum, finite, in a place, in some way corporeal, and almost consisting of matter and form."[235] Now, is this not exactly the premise of Gorlaeus' ontology? Apart from the fact that he rejects the Aristotelian matter-form distinction, is this not precisely how he views God, as a chief representative of the category of *ens per se*?

But there is more. Vorstius' opponents also took exception to his conviction that a rational definition of God could provide a basis of faith and could constitute an instrument of salvation.[236] Faith – they insisted – could not be acquired, any more than salvation. By contrast, Vorstius maintained that reason and understanding could work as instruments of faith, arguing that we may gain insight into God's nature not only through revelation, but also through

> those first and most general principles of a healthy philosophy, which hold no less true in the case of God than they do indubitably hold in general for all other beings, or substances, or spirits, as far as they are based on unchanging foundations.[237]

The closeness of these positions to the premises of Gorlaeus' *Exercitationes* is evident. We recall the latter's definition of philosophy as the "naked knowledge of beings"; his understanding of the *ens per accidens* as a category that includes both God and created things; and his belief that this knowledge of *entia* can provide us with essential knowledge about God and thereby help us perfect our souls. The Vorstian nature of his project is equally evident in his subdivision of his ontological *prima philosophia* into three branches, namely *physica*, *angelographia* and *theosophia*; the latter being defined as the investigation of "the nature and attributes of God" (*de Dei natura et eius attributis*).[238] But *De natura et attributis Dei* also happens to be the subtitle of Vorstius' controversial *Tractatus de Deo*. Although both this theme and the respective description were common, it would have been impossible in the years 1610-1612 not to associated them immediately with the title of Vorstius' book.

At least in the *Exercitationes*, Gorlaeus' intention seems evident. It is to provide the ontological framework from which the truth of Vorstius' specific type of *theosophia* would logically follow. Given that, in the years 1610-1611, he still lacked the theological training and hence the authority to carry out the logical consequences of this ontology for the realm of *theosophia* itself, he limited himself to the parallel case of *physica*, for which his Franeker diploma fully qualified him. In other words, we may conclude that Gorlaeus tried to compose the most ambitious *apologia* possible for the embattled theologian he supported. This tentative conclusion has the advantage of explaining why Gorlaeus decided to leave his *Idea physicae* unpublished and to write his much more ambitious *Exercitationes*, in which his physics would now only feature as the product of an overarching metaphysical synthesis. It might in fact have been Gorlaeus' intention to bring in theology more clearly and prominently in his treatment of the soul. But while he was writing that part of his work, his own soul took to the heavens.

3.11. NICOLAUS TAURELLUS' INFLUENCE ON VORSTIUS AND GORLAEUS

It seems that we have now found a possible explanation for why Gorlaeus may have decided to expand his physical ideas into his more ambitious *Exercitationes* at the height of the battle over Vorstius' theological positions. A comparison of his divine *ens per se* with Vorstius' physicalized God certainly yields a powerful clue as to the genesis and motivation of his ontology. Nevertheless, our puzzle is still not completely solved. We have been specifically looking for the sources of Gorlaeus' atomist convictions. Vorstius' ontology, despite its general concern with the physical interactions between the divine *ens* and the created *entia*, does not rely on any atomist notions.

In order to insert the last piece into our intellectual portrait, we must turn to Nicolaus Taurellus, the author from whom, once again according to Gijsbert Voetius, Gorlaeus had borrowed his doctrine of man as an *ens per accidens*.[239] We recall from above that Voetius mentioned that in 1610, the Heidelberg theologians had sent an evaluation of Vorstius' *Tractatus de Deo* to the Synod of Holland, in which they had established a connection between Taurellus and Vorstius. Let us therefore look at the relevant passage in the Heidelberg report. There, it is stated that Vorstius

> does not hesitate to shake up the doctrine of older and more recent theologians, and he only likes what monstrosities he can find in the gaps of Duns Scotus and in that atheist physician Taurellus: that in his essence, God is a *quantum*, big, finite, composed of essence and accidents, changeable through his will, liable to passive obedience, and three hundred similar things.[240]

We can see that the Heidelberg theologians attributed Vorstius' much-condemned attempts to physicalize God to Taurellus' bad influence. But who was this 'atheist physician'?

Nicolaus Taurellus (Monbéliard, 1547-Nuremberg, 1606), whose name is the Latinized form of Öchslein, studied at Tübingen under Jacob Schegk, to whose independent mode of philosophizing he was to remain attached for the rest of his life.[241] After becoming *magister artium* in 1565, he turned to theology but left this discipline soon for medicine, which he studied in Basel. After becoming *doctor medicinae* in 1570, he obtained a teaching position there as Theodor Zwinger's successor in the chair of ethics. In 1580, he moved to Nuremberg's newly founded University of Altdorf, where he occupied a chair in medicine and natural philosophy until his death in 1606.[242] While Taurellus called himself "professor of Aristotelian philosophy and of Galenic medicine," and published extensively on medical and physical matters, he preferred the title of "Christian philosopher," for throughout his life his true ambition remained the reform of philosophy in such a way that it would suit the needs of Protestant theology.[243]

As he was to recall in bitter autobiographical accounts, he had not left theology for medicine because of any lack of interest, but because he had been repelled by the continuous doctrinal bickering among Protestant theologians and because he was shocked by the customary use made of double-truth arguments, according to which certain statements were clearly false in theology but could be accepted as true in philosophy. Throughout his life, Taurellus not only insisted that truth was one and indivisible, but his most ambitious works were dedicated to the development of a first philosophy that was to provide both philosophers and theologians with first principles and reconcile the Protestant sects. The title of his "Survey of

Aristotelian Metaphysics Emended and Completed According to the Norms of the Christian Religion" signals this objective in a nutshell.[244] That at least some important contemporaries applauded this effort can be gathered from a letter by Rudolph Goclenius, which is prefaced to Taurellus' *De rerum aeternitate* (1604) and which praises both the latter's rejection of the double-truth theory and his concomitant rejection of Aristotle's metaphysics and natural philosophy.[245]

The grand ambitions of this "first Lutheran metaphysician," as Ulrich Leinsle calls Taurellus, expressed themselves at the precocious age of twenty in a bold set of *Theses de philosophia*.[246] These theses were, in a thoroughly modified way, integrated into the large and extravagant *Philosophiae triumphus* (1573), which (in translation) carries the following lengthy and programmatic title:

> The Triumph of Philosophy, that is, a metaphysical method of philosophizing, in which human reasons are thus deduced from divinely instilled ideas that through most solidly constructed demonstrations the truth of the matter will openly shine forth, and Philosophy will burst forth victorious, after having been buried for a long time through the authority of the philosophers. In six hundred questions [on issues] in which Philosophy used to appear to be battling with our revealed truth, but is now reconciled with it, in such a way that she must not only be said to serve faith, but provide its foundation.[247]

It is in this treatise that we encounter the doctrine of man as an *ens per accidens*, which Gorlaeus subsequently adopted into his own philosophy.[248] Why exactly Taurellus insists on the total separability of soul and body is not fully clear and would require in turn a study of the prehistory of his philosophy at the hands of Jacob Schegk. However, a clue might be provided by his desire to attribute the human will exclusively to the soul: "For we are composed of a body and a soul, but the will considered by itself is a simple faculty of the soul, which can understand and exist without the body."[249]

The *Philosophiae triumphus* represents an intelligent, youthful outburst, comparable in spirit with Gorlaeus' precocious system. Like the latter, and despite the occasional applause by individual philosophers and theologians, it met with stern opposition in most quarters. This helps explain why Taurellus, for about twenty years, remained silent on the issue of metaphysics, while in the interval publishing widely on medicine, natural philosophy and emblematics. Still, the replacement of Aristotle's metaphysics remained his secret passion. As Zedler's *Universal-Lexicon* explains in a disarmingly charming phrase, his unwillingness to accept the old metaphysics had to do with the fact that "he could not make Aristotle's doctrine of God, of the intelligences, of providence, and of the soul rhyme [i.e., agree] in any way with the principles of Christian religion."[250]

He shared this aversion to Aristotelian metaphysics with Peter Ramus, but, like Gorlaeus after him, he felt that the solution was not to be sought in dialectics, but in a new ontology.[251] Taurellus had no problem with metaphysics as a discipline; the problem was rather that Aristotelian metaphysics effectively blocked the road to a true knowledge of God.[252] In his eyes, philosophy left to its own devices led to despair, while faith by itself was blind; both were not only in need of each other, but also of a metaphysical 'first philosophy', which could assist them by providing first principles. Like Vorstius and Gorlaeus after him, Taurellus was convinced that a metaphysical definition of being in general, and of God's being in particular, was essential in matters of faith and religion, because it could help us in understanding God and in settling thorny theological questions. An intellectual cognition of God was essential because "who does not know God, will not believe in Christ!" Importantly, 'knowing' for him did not mean any direct acquaintance, but 'ratiocination.'[253]

It is obvious that, very much like Gorlaeus' *prima philosophia*, Taurellus' *metaphysica universalis* had as its goal the discovery, definition, and demonstration of the qualities of being inasmuch as they were shared by all *entia*. Given that it dealt with 'being as being' (*ens ut ens*), this universal metaphysics had to precede all other sciences, including theology.[254]

Having introduced Taurellus, let us now return to the Heidelberg theologians' claim (of 1610) that Vorstius' *Tractatus de Deo* was influenced by this philosopher and physician, and to Voetius' later allegation (of 1641) that Gorlaeus too followed the atheist doctrines of that German philosopher. As to the first link, the extent to which the Heidelberg theologians were correct in discerning in Vorstius' *Tractatus de Deo* the footprints of Taurellus' philosophy is hard to gauge, because although Taurellus is repeatedly cited, his name features far less prominently than those of other authors. It would, for example, be interesting to investigate whether the theologian who most influenced the thought and program of Vorstius' metaphysics and theology was not Girolamo Zanchi (1516-1590). At any rate, Vorstius explicitly mentions as his model Zanchi, who was, like Martin Luther, an Augustinian Hermit turned Protestant.[255] Zanchi, like Vorstius, was animated by 'latitudinarian' ideas, and he had also thought that a good way of settling theological disputes was by defining "the nature of God and the divine attributes."[256] Like Vorstius a few decades later, he too had been attacked for 'physicalizing' God and for causing problems for the Calvinist doctrine of justification. However, in contrast to Vorstius, Zanchi had possessed enough good sense to decline the offer of a Leiden chair in theology. More important than these biographical parallels are, however, the strong overlaps between Zanchi's and Vorstius' respective lists of censured tenets.[257]

As the investigation of the undoubtedly rich background of Vorstius' theology is not our central concern here, we may limit ourselves to the observation that Taurellus and Vorstius share a sufficient number of convictions to explain why the Heidelberg theologians may have perceived them in some manner as allies. Both believed in the rational accessibility of the nature of God and of his attributes, and they both treated God as an *ens* possessing a specific quantity. It is also comprehensible why Taurellus' bold and fascinating *Triumph of Philosophy* might have had a general allure for Vorstius, as it defends man's free will, speaks out against the doctrine of predestination and inveighs against theologians who believe that we, as humans, are merely passive subjects of God's inscrutable decisions – although of course these positions of Taurellus' (and hence Vorstius', to the degree that he followed them) went far beyond anything that either Arminius or the early Remonstrants had ever maintained.[258] On the other hand, the aggressive *Historia crypto-Socianismi Altorfinae quondam Academiae infesti arcana*, in which the heretical connections between the University of Altdorf (where Taurellus taught) westward to Vorstius and eastward to the Polish Socinians are "uncovered" and in which one would expect a confirmation of the claims made by the Heidelberg theologians, remains totally silent on this score.[259] Nor do the very few existing studies on Taurellus explain why the Heidelberg theologians may have reached the conclusion that there was a particularly strong connection between the views of the two authors.[260]

While an exact determination of the nature of this intellectual debt will have to wait until a patient historian of theology decides to tackle this issue, it is an incontestable historical fact that once the Heidelberg theologians had established such a link between Vorstius and Taurellus, everyone else followed suit, not only repeating the original charge but adding further incriminating evidence. The English censors, for example, attributed a number of Vorstius' heretical views to Taurellus' influence; notably the following: that God could be treated "in the predicament of substance," that he possessed not only an essence, but "also accidents" and that he was "somehow limited."[261]

But this process of guilt by association also worked the other way round. If Taurellus was an intellectual enemy of the anti-Remonstrants, he had obviously to be an ally of the Remonstrants! The logic by which one's enemy's enemy is one's friend presumably explains the otherwise inexplicable second edition of Taurellus' *Philosophiae triumphus* in 1617 – in Arnhem, of all places. We have seen above that the person responsible for this edition was most likely Gorlaeus' friend Engelbert Egidius.[262]

It seems evident that our young friend Gorlaeus followed the same reasoning. We have heard above that Taurellus had already been mentioned in De Veno's Franeker lectures, although never in a particularly prominent way. But we may assume that when Gorlaeus learned that this philosopher was the *malin génie* behind Vorstius' theology (about which he must have heard a lot given the presence of several of Vorstius' students at Franeker), he must have tried to get his hands on Taurellus' works. We have specifically suggested above that his friend Engelbert Egidius, whose *Album amicorum* he signed in 1610, may have played a role in the Taurellian turn of his philosophy.

Whatever the exact circumstances may have been, Voetius was certainly right in stating that Taurellus' work exercised a direct influence on Gorlaeus' views. It is very plausible to assume that the latter's plan of going beyond the physics of his *Idea physices* and of constructing an ontology from which to derive theology and philosophy alike was due to Taurellus' *Philosophiae triumphus*.

It is furthermore tempting to view Taurellus' ontology with its atomist implications as the main force behind Gorlaeus' redefinition of Scaliger's minimism in terms of a fully-fledged atomist ontology. Taurellus' ontology generally sheds much light on Gorlaeus' intellectual endeavor. Like the latter, for example, he explains *ens* as a form of the verb *esse* ('to be'), concluding from this that *ens* can only mean 'that which is', and that *esse* and *existere* must therefore be the same. But if, by definition, each *ens* must necessarily exist, Aristotle's characterization of being (*ousía*) has to be just as wrong as the attribution of being to the act and the form of a thing. Nor can there exist such merely potential beings as the prime matter of Aristotelian physics; either prime matter exists *in actu*, or it is no *ens* at all. The next step in the argument states that *esse* is synonymous with being an *ens unum*.[263] Leinsle convincingly argues that by viewing *esse*, *ens* and *essentia* as different only from a grammatical point of view, Taurellus embraces "a nominalist position that is directly opposed to the Thomistic view."[264] But from there, Taurellus takes a further step that is even more relevant for our purposes. If 'oneness' and 'being' are convertible terms, so he argues, then we must conclude that a plurality of *entia* can never merge into a single new *ens*.[265] But this is of course precisely the point of departure for Gorlaeus' own distinction between the atomic *ens per se* and the composite *ens per accidens*! That the *Philosophiae triumphus* is the source of this distinction seems obvious, as it rejects the traditional definition of *ens per accidens* as something that 'is' thanks to the essence of something else, and redefines the term as meaning something that is made up of various essentially unchanged and unchangeable *entia*. Gorlaeus not only adopts this new definition, but, as we have had occasion to see earlier on, also accepts Taurellus' daring conclusion that man must be viewed as just such an *ens per accidens*.

It would not be difficult to list dozens of further traces of Taurellus' *Philosophiae triumphus* in Gorlaeus' *Exercitationes*. Both works, for example, share a dislike for the *ex nihilo nihil* axiom.[266] They equally insist that as all beings exist *in actu*, there is no space for any potency. They both claim that all being, *qua* being, must possess 'quantity' in the sense of having extension and that God too is a *quantum*

– the latter claim being one of those chastised as 'atheistic' by the critics of both Taurellus and Vorstius.[267] In fact, even Gorlaeus' conspicuous 'God criterion' is already in full force in Taurellus: whatever is true of any *ens* in general must also be predicable of God, who after all is the *ens entium*.[268] Indeed, this 'God criterion' in a way constitutes the very core of this theological enterprise: for Taurellus (and for Vorstius, who would busily apply the same idea), to define God ontologically as an *ens* meant to possess an instrument with which to settle disputes concerning such thorny issues as God's ubiquity, prescience or his ability to change his mind. To treat God as a determinate quantity meant, for example, to be able to localize the divine essence and to separate it from terrestrial events, from the consecrated Eucharist or from the actions of men.

Indeed, as Leinsle has put it, Taurellus' "ontology is a theory of existing things and of nothing else."[269] As such, it clearly entails an atomist ontology. The latter necessarily follows from a system according to which (i) all *entia* are actually existing and numerically unique; (ii) all compounds that result from them are mere *composita*, or *entia per accidens*; and (iii) all natural entities, including prime matter, are fully actualized substances.

For this reason, the simplest and most convincing conceptual explanation that we can offer for the genesis of Gorlaeus' ontology is simply this: as he acquainted himself with Taurellus' metaphysics, he must have persuaded himself that in the realm of natural philosophy, he had to transform Scaliger's corpuscular *minima* into proper atoms, and that the latter had to be derived from, and proven by, a theory of being in general.

But here, a curious question emerges. Was Taurellus himself an atomist in the manner of Gorlaeus? Our answer must be a somewhat hesitant 'yes'; for admittedly, we do not find in any of his extant works a clear exposition of physical atomism. But various bits and pieces, when added up, convey a fairly clear atomist picture. His metaphysics of existing units, for example, excludes the divisibility *ad infinitum* that Aristotle demands of all extended magnitudes, which is why Taurellus writes against the Peripatetics:

> The infinite must be sought in number and magnitude. But in the number, it is 'the one'. What about magnitude? There, we have the atom, the point, the indivisibles of motion, the 'now'. To say that all quantity is divisible is an impudent postulate [...] .[270]

Sadly for us, Taurellus refers for his complete proof of the existence of atoms to two works that were either never published or are no longer extant, namely to his commentaries on Hippocrates' *De natura hominis* and on Aristotle's *De lineis insectilibus*; the latter book was even announced in the catalogue of the Leipzig autumn book fair of 1597.[271] In his extant works, by contrast, he treats atoms as if their existence were almost too obvious to require a proof. Given their date of publication, and the scarcity of contemporary philosophers who postulated the existence of physical indivisibles, Taurellus' nonchalance is of course rather surprising.

Unless his 'lost' works resurface, we must remain content with mere hints. There is, for example, the rebuttal of the Aristotelian argument that the indivisible cannot possibly have a size. Taurellus observes that *quantum* and *diairetón* ('divisible') are two separate and independent notions. Being a part of the essence of all being, *quantum* cannot be denied even to the smallest entity. "It is true that atoms are the principles of magnitude," Taurellus states quite casually in his *Kosmologia*, and that everything "is composed out of first, minimal and individual parts."[272] When these 'individual parts' touch each other, they do not become continuous but remain merely contiguous, without merging into a single unity.[273] As for the issue of mixture, his *Philosophiae triumphus* defines it in terms of a mere *compositio* of elements.[274] It may be worth adding here that in his later works, Taurellus also subscribes to Cardano's theory that natural bodies are exclusively composed of earth and water and that mixture occurs under the influence of celestial heat.[275]

Further atomistic explanations are found in his commentary on Arnaldus de Villanova's *Opera omnia*. There he denies that there is such a thing as prime matter, arguing that "it is the common opinion of physicians that there exists nothing prior to, and more simple than, the elements."[276] The limit of divisibility of these elements is not given by any substantial form but by the existence of atoms. This passing reference is highly relevant, for it shows that, like Gorlaeus after him, Taurellus substituted the concept of a natural minimum (which is defined as the lower existential limit of a substantial form) with that of an atom (which designates the absolute size of the basic unit of a substance).[277]

In other passages, however, Taurellus takes his distances from the ancient atomists.[278] This is in fact the last point that must be emphasized here: Taurellus' theologically motivated ontology did indeed entail a doctrine of atomized matter. But both the provenance and the larger aims of this atomism were such that it owed next to nothing to the views attributed to Democritus, Leucippus, Epicurus or Lucretius. Although Taurellus felt that he had to respond to Aristotle's objections to atomism, one can see that it was not his intention to revive the old Democritean model. The premises of this Protestant atomism had indeed little in common with ancient materialism.

3.12. GORLAEUS' CONTRIBUTION TO PHILOSOPHY

As we have just seen, Taurellus furnishes us with the last important piece to the puzzle constituted by Gorlaeus' precocious philosophical system. However, no original thinker is the sum of his predecessors; nor can Gorlaeus' system be deduced from the sum of the views contained in the teaching of De Veno, Scaliger, Vorstius and Taurellus. In fact, only an extraordinarily tidy, coherent and sharp mind could have brought the ideas of such different authors into a new synthesis. This is why Gorlaeus, despite his youth, deserves to obtain a more honorable place in the history of philosophy and science than he has hitherto been granted.

Our knowledge of the circumstances under which the *Excercitationes* and the *Idea physicae* came about certainly do not lessen the respect for the intellectual achievement these two writings represent. At the same time, it obviously influences our interpretation of it. Gorlaeus, as we now know, was not a mature philosopher contemplating physical, chemical and medical evidence, but a self-confident student who responded with his writings to a religious controversy with a number of powerful philosophical notions.

If we resume our findings, we may postulate that Gorlaeus' exposure to specifically Protestant ways of philosophizing began at the University of Franeker. While any Dutch undergraduate in those days was likely to have to come across a range of non-Aristotelian notions, particularly in the fields of metaphysics and logic, Gorlaeus was lucky enough to learn his trade from the versatile and heterodox Henricus de Veno, who in the field of natural philosophy (or *physica*) combined a scriptural approach with German metaphysics and Italian naturalism. Above, we have traced Gorlaeus' conviction that all natural bodies are exclusively mixtures of water and earth and that they are mixed through the force of celestial heat back to De Veno and thence to Cardano and the north-Italian commentary literature on *Meteorology* IV. But we have also seen that Gorlaeus goes beyond his teacher by combining this model with the corpuscular explanations he had found in Scaliger's anti-Cardanian *Exercitationes exotericae*.

Gorlaeus' fascination with Scaliger's work cannot only be inferred from numerous doctrinal parallels, but expresses itself more directly: Scaliger is the only modern author quoted and mentioned by name in all of Gorlaeus' more than 400 pages. However, we have also seen that this admiration is not blind. Gorlaeus' rejection of Aristotelian hylemorphism and its substitution by a fully developed atomist doctrine mark a clear break with Scaliger, who, after all, had depicted himself as the protector of Aristotle's eternally valid physics and had concealed his corpuscularian notions under the Peripatetic terminology of *minima naturalia*. To explain the sources of Gorlaeus' atomist ontology, we have had to look elsewhere. If the arguments presented in this chapter are correct, then his ontology and the atomism that results from it is not the fruit of Italian medico-philosophical physics, but of German theologico-metaphysical thought.

We have furthermore seen how Gorlaeus, as a beginning student of theology at Leiden University, found himself at the epicenter of the Arminian controversy and the upheaval caused by the appointment of the theologian Conrad Vorstius as Arminius' successor. Everything we have heard about Gorlaeus – his teacher De Veno, his friend Engelbert Egidius, his roommate Rudolph van Echten, and above all, the doctrinal overlaps with Taurellus and Vorstius – document that he sided with the embattled professor. His propagation, in the *Exercitationes* of a 'first philosophy', ontologically defined, as a method for finding the essential properties of all existing things, including God, can in the context of 1611 be interpreted as nothing else but a defense of the guiding idea behind Vorstius' controversial *Tractatus de Deo*, which had been published in its final form in 1610.

We have heard that one of the charges brought against Vorstius was that he had followed the 'atheist physician' Nicolaus Taurellus in applying physical categories to God. The Utrecht theologian Voetius was later to accuse Gorlaeus of having committed the same crime. This charge, we have found, is correct. We have shown that Taurellus' ontology provides a number of crucial elements for our reconstruction of Gorlaeus' atomism, for it represents the blueprint of the latter's equation of 'being' with 'existence', 'oneness' and 'quantity' – an identification from which atomism follows as a corollary. Put somewhat crudely, Gorlaeus accepted Taurellus' atomist ontology and applied it to his Italianate natural philosophy, thereby producing a philosophical and physical system that was remarkably original.

That system, as we shall see in our concluding chapter, was for some decades read, discussed, accepted or refuted at home and abroad, leaving its most visible and intriguing traces notably in the circle of Descartes' Dutch friends.

Chapter 4

Gorlæus' Place in the History of Seventeenth-Century Thought

Gorlaeus died in the spring of 1612. We do not know where and why he died, nor whether he had been ill for some time before passing away; perhaps he had contracted malaria, as presumably did his father, who also died in 1612.[1] All we have is his tomb and his two posthumous publications, the *Exercitationes philosophicae* (published in 1620) and the *Idea physicae* (published in 1651). From the unfinished state of the concluding part of the *Exercitationes*, we may conclude that Gorlaeus was still working on his longer treatise when he died. One in fact senses that his decision to deviate from his standard type of exposition and to jot down a paraphrase of a number of positions defended in a recent theological disputation at Leiden concerning the origin of the soul and the transmission of evil must have been due to his wish to conclude his manuscript before it was too late.[2]

As we have seen above, the *Idea*, printed in 1651, contained ideas that were too similar to those expounded in the *Exercitationes* and appeared too late to elicit much of a reaction.[3] With the *Exercitationes*, the story is different: it enjoyed, notably between 1620 and 1650, quite a reputation at home and abroad.

Who decided to publish the *Exercitationes* in 1620, eight years after Gorlaeus' death, is not clear. After all, by then, Gorlaeus' parents had also passed away, and Gorlaeus' aunt and his cousin Carel van Gelder, a lawyer, were taking care of the estate. It might have been Van Gelder, the executioner of the will of David van Goorle, Sr., who had the book published. One could also imagine an involvement of Abraham van Goorle, Jr., David Gorlaeus' paternal cousin. Jaeger has suggested that Petrus Bertius, the former regent of the Theological College at Leiden as well as a friend of both Arminius and Gorlaeus' uncle Abraham, may have organized this publication – but we have already seen that there is no evidence to corroborate that claim.[4] What is conspicuous, however, is that whoever it was who edited the work had no interest in revealing himself or his motivations. Contrary to the *Idea physicae* of 1651, the *Exercitationes* do not carry any preface or introduction. Whether this absence of any explanation points to an awareness of the doctrinal incompatibility of its contents with the contra-Remonstrant orthodoxy that had been imposed on the entire country in 1619 at the Synod of Dort cannot be determined, although it is likely.

Once, however, that the manuscript had been printed and released, it entered the public domain and started, like any other book, to lead a life of its own. As has been shown in chapter 1, the life of this particular book was even more independent of the author's intentions than most other books, as none of its readers, with the exception of a very few, could associate the author's name with any definite social status, ideological or religious stance or the vaguest biographical data.

This last chapter is therefore dedicated to the reception of Gorlaeus' work in the seventeenth century. It is of course impossible to provide a complete *Wirkungsgeschichte* in a few pages. What will be offered, by contrast, is a rough sketch of the way in which Gorlaeus' ideas were absorbed into, and thereby influenced, seventeenth-century thought in the domains of philosophy and science.

4.1. INTERNATIONAL RESPONSES TO GORLAEUS: THE PARISIAN CASE

A valid idea of the fortune of Gorlaeus' *Exercitationes* and its international reception can be obtained by examining a circumscribed case study. Paris, one of the early seventeenth century's most important European centers of learning, offers itself here. For this reason, we shall now turn to examine the role Gorlaeus' *Exercitationes* played in a series of Parisian authors in the years 1620-1650.

We recall from chapter 1 that seventeenth-century readers tended to view Gorlaeus as a *novator* – a category invented to capture those philosophers who, unhappy with the prevailing Aristotelian system, had proposed an alternative to it. Contrary to other categories, this one was accepted by friends and foes alike, who used it either as a term of endearment or of abuse. To those who had grown weary of the received philosophy of the schools, the *novator* was a welcome harbinger of hope. To those, by contrast, who continued to uphold Aristotle as the measure of all solid thought, the *novator* constituted a nuisance.

It has been mentioned in chapter 1 that Marin Mersenne (1588-1648) repeatedly referred to Gorlaeus in the period in which he still behaved as a verbose and polemical soldier of the *ecclesia militans*; that is, before his transformation into the irenic Secretary of the Republic of Letters. In those days, he regarded Gorlaeus as one of the authors who had to be combated and condemned. In his commentary on Genesis (1623), he listed him alongside Campanella, Bruno, Telesio, Kepler, Galileo, Gilbert, Bacon, Fludd and Hill – most of these philosophers being either Protestants or else Catholics who had been convicted, condemned or even burnt by the Inquisition.[5] In *La vérité des sciences* (1625), Gorlaeus is mentioned in the company of Bodin, Carpenter, Hill, Olivi, "and various others."[6] In *L'impiété des déistes* (1624), finally, Mersenne provided a sketch of an encyclopedia in which he intended to refute all kinds of lies, and notably those of "Gorlaeus, Carpenter, Basson, Hill, Campanella, Bruno, Vanini and several others."[7] The announced encyclopedia was, unfortunately, never published. However, Mersenne does hint at what he finds "impertinent" about these authors. He laments Carpenter's and Gorlaeus' subscription to the principle that "all things are made and derived from nothing," Gorlaeus' and Hill's atomism and notably the view "that inside bodies there are atoms which have quantity and figure."[8]

But not everyone in the Paris of the 1620s disliked the *novatores*. For the same reasons that Mersenne attacked him, Gabriel Naudé (1600-1653), an influential librarian and scholar, seems to have liked him. In his well-known princely *Advice on Establishing a Library* of 1627, Naudé (who was repeatedly asked to set up learned libraries) dedicates a specific chapter to the organizing principles of a large library as required in an age of knowledge proliferation. After speaking of theology (for which he proposes a systematic ordering of the books), he turns to philosophy, a discipline for which he deems a chronological set-up to be the most effective method of organization:

> In Philosophy, one has to start with that of Hermes Trismegistus, which is the oldest philosophy, continue with Plato's, Aristotle's, Ramon Lull's and Ramus', and finish with the *novatores* Telesius, Patrizi, Campanella, Bacon, Gilbert, Bruno, Gassendi, Basson, Gomez, Carpenter, and Gorlaeus, who are the most important authors in a flood of others.[9]

In their different ways, Mersenne and Naudé document the fact that Gorlaeus was known and read in Paris already in the 1620s, and was considered an important figure both by traditionalists and *libertins*. Maybe it was Hugo Grotius (1583-1645) who brought Gorlaeus to Paris, where he was a regular visitor to the Cabinet Dupuy. At any rate, Gorlaeus' reputation continued throughout the century. Take, for example, Jean Bachout, who, in 1651, translated and introduced Jean d'Espagnet's *Enchyridion physicae restitutae* (1623). In his *Discours*, which recommended the *Enchyridion* and at the same time revealed its author's hitherto anonymous identity, Bachout invoked the idea, much en vogue in those years, that Aristotle had for centuries managed to remain a kind of philosophical dictator only because he had wilfully obfuscated the more noble and ancient philosophies of the Egyptians, Chaldeans, Presocratics and even of his own teacher Plato, whose myths still contained some of the ancient splendor of truth. Fortunately, Bachout continued, recent times had seen the rising of valiant men who boldly sought to recover the hidden truth. Italy had begun in this enterprise thanks to Telesio, Patrizi and Campanella. But then, other countries had joined in:

> Germany and England, too, have had several men who only followed Aristotle's opinions in those places where they found them at their most reasonable, as did Bacon, Fludd, Gorlaeus, Taurellus, Carpenter and others, of whom some have proposed new principles.[10]

As for France, so Bachout concluded his introduction, she had offered above all three men to this enterprise: Ramus first, subsequently d'Espagnet (to whose very book Bachout's *Discours* is prefaced), and finally Descartes. Notable about this list is not only Gorlaeus' appearance among these important new philosophers, but also the fact that Taurellus (whom we otherwise never encounter in such lists of *novatores*) is placed next to him. It is evident that Bachout must have been aware of the Utrecht Crisis of 1641, in which the names of both Gorlaeus and Taurellus were invoked in the context of Cartesianism.

Obviously, it is one thing to be aware of Gorlaeus' name and work, and to place his *Exercitationes* in its appropriate slot within a princely library, and quite another to use him and to apply his ideas. That his book was actually read and used can be shown by our fourth French example, that of Charles Sorel (1582?-1674). Among the numerous works written by this novelist and polymath, the four parts of his *Science universelle* (published in 1634, 1637, 1641 and 1644) combine Baconian themes such as the methodology of (collective) scientific experimentation and the perfectibility of man with topics and ideas taken from other new philosophers, and notably from the atomists among them. In this particular work, as well as several others, we encounter small chapters containing, as it were, short doxographies of the *novatores*. It is not only the encyclopedic inclination of the polymath that led Sorel to write such chapters.[11] Adapting a Baconian model for his own ends, he also treated the thoughts of others as the quarry in which to find the materials for the erection of his own natural philosophy. In his doxographies, he illustrated the way in which *novatores* had often mixed truth with fancy in their desire to do away with Aristotle, but nevertheless deserved our applause for having opened the doors to a new, empirical and much more fruitful approach to the investigation of nature. One of his shorter chronologies reads like this:

> One must praise the great courage of Telesio, for having been the first to dare to censure the ancient errors. [...] Patrizi must also be strongly commended for having enlightened his age with respect to numerous absurd views concerning celestial and terrestrial bodies. These men, together with Cardano and some others [...] have revealed that the true number of the elements is only two, [...] a view with which Gorlaeus and the author of the *Enchyridion* [d'Espagnet] agree. Copernicus, Galileo, Giordano Bruno and Descartes teach us all that we can imagine and suppose with respect to the number, situation and the movement of the principal bodies of the universe.[12]

In this short survey, Gorlaeus is mentioned merely as one among those who subscribed to the sixteenth-century Italian naturalists' limitation of the number of elements. A more detailed appraisal is found in Sorel's *De la perfection de l'homme* (1655), which contains a long section dedicated to the history of the new philosophy and opens with a long praise of the *novatores*. Toying with a Baconian motif, Sorel insists that at least since the discovery of the New World, it should have been obvious that there are more secrets to nature than had been fathomed by the Greeks. Empirical research was therefore necessary. "Although the very name '*novator*' is loathsome to a number of people, one has to take into consideration that whereas one has to worry about this phenomenon in matters of theology, this is not the case in natural and human philosophy."[13] Once again, Sorel lets his history begin with the Italian naturalists Telesio, Patrizi and Cardano. After a very thoughtful and partly critical evaluation of Ramus' contributions, Sorel turns to astronomy (Copernicus, Tycho, Kepler and Galileo), and adds a separate and surprisingly positive evaluation of Giordano Bruno, whom he casts as an 'astronomical novator', omitting his metaphysical and theological ideas.[14] After speaking of Bernard Palissy's empirical work on chemical substances, Sorel turns his attention to Gorlaeus, who receives a section on his own:

> Following a chronological order, we shall now come to *novatores* who have written in Latin and some others who have preceded them and who have followed the rules of philosophy. Among the moderns who deserve to be mentioned here, there is a certain David Gorlaeus, a Dutchman, who has written a book called *Exercitationes philosophicae*, where he engages in a fight against the entire theoretical philosophy of the Peripatetics. After dealing with metaphysics, he turns to physics. He treats all corporeal qualities, proposing a number of views of which some are genuinely new, while others are simply proposed anew. He shows that what one calls Heaven is nothing but the extension of air, and that there are only two elements, earth and water, and that fire is no element at all, but a simple accident. One cannot entirely grant him this point, because fire is only to be esteemed to be an accident with respect to the fire which we artificially produce. One must recognize that there is a different type of fire which is a veritable substance, which, if it is not an element, must at least be taken to be one of the principal bodies constituting the world.[15]

Once again, Sorel limits himself to mentioning Gorlaeus' two-element theory, omitting his metaphysics – although it is there that he would have found the foundations for an atomist conception that, when he finds it in other authors, tends to enthuse him, but about which he remains strangely silent in this particular case. Moreover, although he applauds Gorlaeus for being a *novator* in some respects and a *renovator* in others, he rejects the suggestion that real, natural fire is merely an accident of the other elements. While he does not belabor this point, both the role played by the sun and by celestial heat in his own system, and our own critical remarks about the odd status that the real accident of heat plays in Gorlaeus' natural system, render Sorel's caveat quite comprehensible.

In the light of this criticism, one might wonder why Sorel should have decided to dedicate a separate section of his doxography to Gorlaeus at all. The answer, I think, lies hidden in Sorel's earlier work, and notably in his exhaustingly prolix *Science des choses corporelles* of 1634, whose programmatic subtitle almost echoes Gorlaeus': "où l'on connoist la Verité de toutes les choses du Monde par les forces de la Raison et l'on trouve la refutation des Erreurs de la Philosophie vulgaire." Although in this particular treatise, Sorel cites no recent philosopher by name, his extended discussion of the elements resembles Gorlaeus' reasoning so strongly that it is hard to reject the idea that Sorel wrote his own treatise with Gorlaeus' close at hand. Take, for example, his lengthy chapter XII, "On the number of the elements," which ends, as it were, with two summaries. The first reads as follows:

> We have thus found that earth, water, air and fire are four different bodies, but they are neither the four elements nor all simple bodies. Earth, water and air are thought to be simple, while fire is a composite. Earth and water serve for the composition of all bodies, without air being a part of their substance even though it is sometimes found among them. Fire is also just the same air, when heated, and must be counted among the first bodies only inasmuch as it is an agent, but is no part of the composite.[16]

The overlap with Gorlaeus' theory of the elements is obviously very strong, but – one might object – so is their joint debt to Cardano. Gorlaeus' specific influence on Sorel becomes manifest however in the latter's attempt to reduce his two-plus-one theory of the elements to an atomistic theory of the dry and the wet, whereby warm and cold are described as mere modifications of these two essential properties.[17]

Sorel provides us with a good example of the type of influence that the *Exercitationes* managed to exercise in the period 1620-1650 in certain quarters. In Sorel's view, Gorlaeus' two-element theory was closer to practice and experience than the transmutationalist four-element theory of the Aristotelian tradition. "One has to make tests," Sorel wrote in 1634, "to find out how many simple bodies make up mixtures." What tests did he have in mind?

> If you press plants or the flowers and the fruits of trees, water will ooze out, and what is earthen will remain. If the flesh of an animal is sliced up, the blood and

> the humors will flow out, and what is solid will remain separated. Here, then, you have water and earth; but where is the air and the fire?[18]

Such commonsensical observations are obviously neither sophisticated nor irrefutable, and they clearly do not add up to a 'test'. But fortunately, it is not the early Sorel's experimental methodology that is at stake here, but rather the role that Gorlaeus' theory of the elements plays in it. What emerges is that Sorel, who wished to put Baconian experimentalism, an improved theory of matter and a new cosmology under the yoke of a new science of his own devising, viewed Gorlaeus' atomistic theory of the elements as a fruitful contribution to this program. It is typical for Sorel, and *mutandis mutatis* for other seventeenth-century philosophers too, that Gorlaeus' importance diminished once Descartes had published his work; this explains his diminished role in Sorel's later doxographical chapters. Not that Sorel was entirely convinced by Descartes' picturesque natural philosophy. He did, however, consider him the boldest of the *novatores*, and the most wide-ranging: "Among the *novatores*, one finds none that has removed himself farther from common thought," he explained. Moreover, Descartes' philosophy was all the more persuasive as he illustrated it with "paintings" and "pictures [...] full of these little bodies, which are so little known, but which are represented there with such assurance as if he had seen them clearly."[19]

The 1655 version of Sorel's doxography of the new philosophy culminated with Descartes' natural philosophy, on the one hand, and with the recent triumphs of chemistry and astronomy, on the other. Gorlaeus' natural philosophy, while retaining an honorable place in the gallery of the *novatores*, was by then slowly turning into a mere precursor of greater and more radical models of nature.

4.2. DUTCH RESPONSES TO GORLAEUS AND THE RISE OF CARTESIANISM

The Parisian case study shows the role that was attributed to Gorlaeus' *Exercitationes* in the international panorama of the *novatores*. But let us return to the Netherlands and examine the influence of his thought in his country of origin, and in particular the real or perceived link between his own thought and that of Descartes. Although nowadays, only two copies of Gorlaeus' *Exercitationes* can be found in Dutch public libraries, this book was at the time, in Theo Verbeek's words, "in everybody's hands."[20] One finds it used in university lectures, discussed in academic disputations and referred to in the philosophical literature. Even a philosophy professor such as Adriaan Hereboord, whose own philosophy took a very different direction, did not hesitate to recommend Gorlaeus' metaphysics in his "Advice on the method of studying philosophy" of 1648.[21]

It is particularly noteworthy that a series of leading philosophers, including Isaac Beeckman, Henri Reneri, Henricus Regius and Jacob Ravensperger, who in one way or another are all linked to Descartes' years in the Netherlands, owned or used the *Exercitationes*. The intriguing possibility of its influence, not only on these Dutch thinkers, but on Descartes himself, therefore emerges with force.

The wide academic diffusion of Gorlaeus' *Exercitationes* in the Netherlands is shown by the use made of it by Jacob Ravensperger (or Ravensberg, 1615/6-1650). Interestingly, Ravensperger was the son of a theologian from Steinfurt who in 1614 had been appointed as the University of Groningen's first professor of theology and whose lot in the Netherlands was decidedly more fortunate than that of his Steinfurt colleague Vorstius at Leiden. His son, Jacob, studied in Groningen, where he took his degree in philosophy in 1639. In 1641, he became professor of mathematics at Utrecht University. It seems that he tried to maintain an equidistant attitude towards the two quarreling camps during the Utrecht Crisis (1641-43); although incidentally, Regius' disputation that ignited it all happened to be dedicated to Ravensperger as well as to the Utrecht theologians. His own extant student disputations show an open-minded attitude. For example, he did not hesitate to cite Descartes' law of refraction and vortex theory. In 1648, his appointment was enlarged to include physics. In 1650, however, he died, at the age of 35. In his funeral oration, his colleague Daniel Berckringer praised not so much his genius or eloquence as his wide-ranging curiosity and his vast erudition, which spanned physics, mathematics and medicine, and reached peaks in metaphysics and scholastic speculations.[22]

Scholars have worried about the question of whether Ravensperger should be labeled an Aristotelian or a Cartesian in the context of the Utrecht landscape. Once again, however, it should be remembered that philosophy in the Netherlands had progressed far beyond Aristotle by the 1630s and was influenced by a host of non-scholastic authors. To demonstrate this point, let us take a look at the disputation *pro gradu* (that is, for his philosophy diploma) that Ravensperger defended at Groningen in 1639. In its 85 theses, he discussed an impressive array of issues ranging from logic and metaphysics to physics, arithmetic, geometry, astronomy, geography, optics and finally ethics and politics; addressing such different issues as the structure of ontology, the number of elements, the epistemological reliability of the telescope and the aristocratic status of the equestrian families of the Roman Empire. In addressing these problems, he quotes an impressive range of ancient and medieval authors (Plato, Aristotle, Seneca, Lucretius, Aquinas, Scotus, Aureoli, etc.), Catholic and Protestant scholastic authors in the domains of philosophy and theology (Zabarella, Toletus, Fonseca, Suárez, Keckermann, Goclenius, Timpler, Burgersdijk, Calvin, Piscator, Bellarmine and many others), but also more modern, 'scientific' voices (including Copernicus, Kepler, Rheticus, Ramus,

Alsted, Fludd, Commandino, Bacon, Gilbert, Sennert, Lansbergen, and Galileo). Gilbert and Galileo are in fact described as "heroes of our century."[23]

What is of interest in the present context is that among the *novatores*, we repeatedly encounter also our Gorlaeus, whose *Exercitationes philosophicae* is treated as an authoritative text, although Ravensperger disagrees with it as often as he agrees. On the question of "Whether there exists only one theoretical science," for example, Ravensperger sides with Aristotle and Seneca, who denied such a singularity, against Gorlaeus and Antonio Bernardi della Mirandola (1502-1565).[24] On the question of "Whether there exists an objective, single and adequate concept of being," which he thinks we may affirm, he rejects the view of the nominalists, Thomists and "some recent authors, such as Johannes Combach and David Gorlaeus."[25] With respect to the question of "Whether fire and air are elements," Ravensperger takes no position between the "the common and Peripatetic view," which affirms it, and the negationist positions of Cardano, Timpler and Gorlaeus: "Choose what you wish!"[26] In the subsequent question, on "Whether air is by nature neither hot nor cold," the disputant openly sides with Gorlaeus, "affirming that air is no element."[27] Turning to arithmetic, where Ravensperger raises the question of whether material, quantitative units and numbers should be comprised in the category of transcendental number and unity or are possibly even identical with it, we encounter Gorlaeus once again, among the authors who takes the second, more radical view.[28]

Two aspects of Ravensperger's use of Gorlaeus deserve to be mentioned here. First, it is noteworthy that, in contrast to the Parisian authors we have just examined, his interest is not limited to Gorlaeus' two-element theory. As his references document, he has examined the entire *Exercitationes*, extracting from it useful theses for his own general philosophy, metaphysics, physics and arithmetic. What is curious, of course, is the absence of any interest, on Ravensperger's part, in Gorlaeus' atomistic ontology. The second point that must be mentioned here is the nonchalant self-evidence with which Gorlaeus, the first-year theology student, appears in the company of authorities whose name recognition has remained intact. It is obvious that in the 1630s, Gorlaeus was an author who was not only widely read but also used at Dutch universities. A detailed examination of the dissemination of his ideas remains, however, a desideratum.

In 1641, with his doctoral title from Groningen in hand, Ravensperger moved to Utrecht University where he would teach mathematics and later also physics, and where he arrived just in time to witness the explosion of the Utrecht Crisis surrounding Descartes and his teaching. In this Crisis, Gorlaeus would once more play an important part. Let us therefore move – together with Ravensperger, as it were – to Utrecht, and begin by taking a look at Henricus Reneri (1593-1639), Utrecht's first professor of philosophy. Reneri has variously been described as "the first among the Batavians whom Descartes got to know," which is false, since the latter's friendship with Isaac Beeckman began ten years earlier; or as "the first Cartesian," which is also somewhat inaccurate given that Reneri was not only three years older than his alleged master, but, as his extant publications indicate, also combined Cartesian ideas with a range of ideas of a different provenance.[29] What linked Reneri and Descartes was certainly a profound friendship, a common native language and a number of shared ideals. But clearly, the relationship was not symmetrical, as Reneri's admiration for Descartes was boundless.[30]

Reneri, originally a Catholic Walloon, had first been trained in philosophy in Louvain and then in theology in Liège, where his reading of Jean Calvin's *Institutio Christianae religionis* provoked his conversion to Protestantism. He therefore took to the Netherlands, where he enrolled as a student of theology in Leiden in 1616; that is, five years after Gorlaeus. In the years to come, we encounter him in the various roles of private tutor, student of medicine and mathematician. At the same time, we also know that he was engaged in chemical, optical and meteorological experiments and made inventions in the fields of thermometry and optics. In 1628 or early 1629, he encountered Descartes, whom he befriended; so much so that between May 1632 and February 1634, Descartes moved to Deventer, the city where Reneri was teaching at the time. But Descartes was not the only Frenchman who encouraged his scientific activities and influenced its methodology and theoretical underpinnings. In the same period in which he first met Descartes, Reneri also encountered Pierre Gassendi, who was touring the Low Countries in 1628-29.[31]

Descartes and Reneri continued their close collaboration after Reneri had moved to Utrecht in 1634, where he taught philosophy at the Illustrious School, which was upgraded to a full university in 1636. Once again, Descartes followed what appeared to have been at the time his most important contact in the Dutch university world. We know that Reneri took lessons in mathematics from Descartes while helping him with the distribution of the *Discours de la Méthode*. He also gave private lessons in which he used Descartes' *Discours* and the *Essais* that were attached to it. In a letter to Mersenne, Reneri wrote that Descartes "is my light, my sun, and what Vergil said in his *Bucolics*, I can say about him: 'He will be for me forever a god.'"[32] Remarkably enough, when he died in 1639, the funeral oration that his colleague Antonius Aemilius delivered, spoke more about Descartes and the promises of his philosophy than about the deceased Reneri. "In short," Verbeek concludes, Reneri seems to have been "the main herald of Descartes' glory."[33]

And yet, when one reads Reneri's inaugural lecture of 1634 and his extant university disputations, one finds that they contain remarkably few Cartesian elements. What one encounters instead is a combination of an empirical, indeed experimentalist, attitude that owes perhaps more to Bacon than to Descartes, with philosophical notions that are taken from a plethora of sources.[34] In fact – and with this we return to our present theme – from his disputations, Gorlaeus is precisely one of these non-Cartesian sources. Although it is impossible to offer a water-tight proof of this line of influence, in light of the fact that Reneri's disputations almost never cite the authors whose views are being used, we cannot only point to doctrinal parallels but fortunately we also know that Reneri owned a copy of Gorlaeus' *Exercitationes*.[35]

The way in which Reneri absorbed Gorlaeus' ideas into his university teaching can be exemplified by the disputation *On the Elements*, which Reneri had a Hungarian student of his defend in 1635. Gorlaeus' specific imprint is recognizable from the very beginning, as Reneri rejects the traditional distinction between elements and mixts, arguing that this distinction does not do justice to "air, which is neither an element nor a mixed body constituted by the elements."[36] But as the disputation proceeds, we quickly discern other influences too. The traditional view of the elements is also false, Reneri has his student explain, because the assumption that mixts can be resolved into these elements is contradicted by the fact that gold cannot be resolved into anything else by nature or by art. From this Reneri concludes that "a connection between the elements can be produced such that their link can afterwards no longer be resolved by natural causes."[37] This view, which owes much to iatrochemistry and entails a molecular understanding of matter, is, in turn, combined with an atomist definition of elements. The traditional division of all bodies, including elements, into prime matter and form is unnecessary, as "no phenomenon is observed in the elements that requires more than matter and its diverse dispositions with respect to quantity, figure, motion and quiet."[38]

Reneri then returns specifically to the question of the number of elements. The number four, he explains, has of course always had a great appeal: the combination of the four qualities are said to lead to the four elements; which, in turn, are linked to the four temperaments, humors, ages and seasons. Reneri shows, however, that this entire model of quaternaries relies on circular reasoning: one supposes, for example, that food is made up of the four elements in order to demonstrate that our body is also composed of the same number of elements.[39] He also makes short shrift of the Aristotelian theory of gravity and levity, arguing that air is always heavy, even if we do not feel it (just as fish do not feel the weight of water).

Reneri's corpuscular conception of the elements – which combines ideas found in ancient sources with notions found in Gorlaeus, Gassendi and Descartes' *Météores* – becomes ever more obvious as the disputation progresses. We hear that elements do not transmute, but have been generated entire and directly by God (theses 17-18). Moreover, each element possesses a specific figure (thesis 14), just as sensory experiences are caused by the figures of particles impinging on our organs. Following the model provided in Plato's *Timaeus*, which explains the burning taste of certain foods by the puncturing of the tongue by pyramidal fire particles, but at the same time subverting it, Reneri explains that the heating and burning quality of certain liquors is not due to the presence of fire but to "certain figures and the smallness and mobility of the minimal parts of which these bodies consist," which provoke a feeling "similar to that of fire."[40] The reason Reneri rejects Plato's fire particles is because he rejects the notion that fire is an element. Air is not an element either. In arguing this point, Reneri would seem to rely once more on arguments provided in Gorlaeus' *Exercitationes*: although air is found everywhere and fills the sky as well as all otherwise unfilled pores, it is not part of any mixture. However, Reneri adds a morphological (Democritean, Platonic, Gassendist, or indeed Cartesian) twist to this two-element theory: Earth and water, he suggests, cohere because they have fatter parts, while air is too fluid and subtle to attach itself to them.[41] In his hands, mixture has become a matter of cohesion, and cohesion, in turn, a matter of atomic shapes. And while this definition of mixture goes beyond Gorlaeus' (who, as we recall, did not explain the ability of elements to enter into mixture by means of their atomic shapes, but by other qualities), his two-matter theory remains faithful to the latter's model: "Water and earth, considered in their purity, are the only elements properly so called, because they are simple and because one has to recur to them to explain the generation of all mixts and the resolution of anybody whatsoever into them."[42]

Of both Reneri and Regius – Descartes' earliest supporters at Dutch universities – it is often said that they were fascinated by Descartes' physics, but did not follow him in his metaphysical evolution. Reneri's "lack of interest in metaphysics" and his unwillingness "to defend Descartes' metaphysics" have been contrasted with his enthusiasm for the physical theories of his much-admired friend.[43] By the same token, Descartes' disgust over Regius' *Fundamenta physices* (1646), which presented a Cartesian natural philosophy without its metaphysical foundations, is famous as it was aired not only in vitriolic letters to his erstwhile friend, but also in Descartes' preface to the French translation of his *Principia*. Against this apparently anti-metaphysical background, it is all the more surprising that Gorlaeus' thesis of man as a composite being, an *ens per accidens*, should have made its appearance in a thesis defended under Regius.

Let us therefore turn to this second early Dutch friend of Descartes'.[44] Henricus Regius (1598-1679), who was born in Utrecht six years after Gorlaeus, had, like him, taken his first diploma at Franeker University. He proceeded to study medicine in Groningen, Leiden and Montepellier. He eventually took his doctorate in Padua, having studied with such celebrities as Santorio Santorii and Cesare Cremonini. Upon his return to the Netherlands, Regius first worked as Utrecht's town physician, then moved to Naarden, where he directed the Latin School, but also ran into difficulties with the local church authorities, who found his behavior unorthodox. His initial unwillingness to sign, as his function demanded, the Reformed Act of Faith led to a visitation by members of the Amsterdam Classis of the Reformed Church, who in their report concluded that he was poisoned by Arminian and Socinian heresies.[45] Although his skirmish with the Church came to an end after he apologized and signed the Act of Faith, suspicions of heterodoxy resurfaced a few years later.

In 1634, Regius returned to Utrecht, where he gave private lessons and resumed his duties as town physician; in 1638 he was appointed professor of theoretical medicine and botany. In the same year, his colleague Reneri introduced him to Descartes, whose *Essays* of 1637 he had admired and incorporated into his own private physics lessons and with whom he now initiated an intense correspondence. In his first, introductory letter to Descartes, Regius in fact suggested that it had been his private physics lessons, inspired by Descartes' *Essays*, which had led to his appointment as university professor. Because of his keen interest in natural philosophy, the university asked him in 1640 to lecture additionally on natural philosophy, but given the presence of a more senior professor of philosophy at Utrecht University, Arnold Senguerd, this appointment was due to lead to frictions.

When Regius wished to expose in public his thoughts on natural philosophy, which he seems to have written down in a *Compendium*, Descartes advised him against publishing such a book (possibly, Verbeek suggests, out of fear "that his new friend might forestall him"), proposing a disputational format instead. By contrast, Utrecht University's newly appointed rector, the theologian Gijsbert Voetius, who worried about tensions between rivaling philosophers, first advised him to dispense with publicizing his physics altogether, but in the end conceded to him the possibility of packaging his natural philosophy in medical disputations.[46] Neither Regius nor Descartes nor Voetius could have foreseen that these disputations would lead to the stormy affair that would culminate in the public prohibition of Cartesian ideas at Utrecht University.

All of Regius' disputations of 1641 dealt in one way or another with *physiologia*, a term that could possess a more narrow medical definition (thanks to Jean Fernel's homonymous treatise of 1542), but which could also stand in as an equivalent of both *physica* and *philosophia naturalis*, being defined as a general theory of the operations of natural bodies.[47] In other words, the polyvalent term 'physiology' allowed Regius to stay either within the boundaries of medicine or to foray into the territory of natural philosophy. In his first set of disputations, which he published under the overall title *Physiologia sive cognitio sanitatis*, he remained by and large within the domain of medicine. But in the second set, which was published under the overall title *De illustribus aliquot quaestionibus physiologicis*, he clearly entered the fields of physics, metaphysics and, in the perception of his contemporaries, theology, triggering what would become the Utrecht Crisis.

It is known that Descartes followed the composition of these disputations closely, suggesting various improvements.[48] It is thus not surprising that the disputations have a Cartesian ring to them. Some doctrines are directly and consciously taken from the French philosopher. Citing the *Dioptrics*, for example, Regius reproduces Descartes' doctrine concerning the role of the pineal gland in the transmission to the soul of impressions conveyed by nerve fibers.[49] His theory of matter seems also very much aligned with Descartes': "While other physicians and philosophers have traditionally paid little attention to those insensible particles," Regius writes with reference to those microscopically small corpuscles that defy sensory perception, "we believe that innumerable natural mysteries depend on them."[50] Indeed, his entire theory of matter appears to correspond to Descartes': apart from the mind, which is an entity by itself, all material bodies can be reduced to variously shaped particles of a universal matter, whose motions and constellations explain physical properties. To help students memorize this doctrine, Regius composed a couplet, which sounds like an early shorthand manifesto of what Robert Boyle would later view as the program of the "mechanical philosophy":

> Mind, measure, quiet, motion, position and figure
> Are together with matter the origin of all things.[51]

This distich was amply ridiculed by Martin Schoock, one of Descartes' fiercest detractors, as a kind of senseless prophetic oracle (although, as Erik-Jan Bos has document, it remained popular among university teachers and students well into the eighteenth century).[52] But while for Schoock it was evident that Regius was nothing but Descartes' mouthpiece, Descartes himself was aware of notable differences between Regius' views and his own – differences that would eventually lead to their complete rupture.

At the same time, it should be remembered that, in the face of all the admiration that he nurtured for Descartes, Regius was an independent thinker who had lectured on physical and medical issues before he read Descartes' *Essays* in 1637 and became acquainted with their author the year thereafter, at the mature age of 40. In other words, unlike later Cartesians, he did not begin philosophizing on the basis of Descartes' work. Among the various other authors who influenced his thought was Gorlaeus, whose *Exercitationes* Regius had studied before reading any of Descartes' works. For example, several of his positions sketched in 1640 in his correspondence with Descartes are so close to Gorlaeus' ideas that we may suppose that he borrowed more doctrines from the *Exercitationes* than just the definition of man as an *ens per accidens*.[53]

So, if in his physiological disputations, he rejected substantial forms and posited microscopically small particles instead, we need not automatically conclude that he had acquired these views only recently and that they were exclusively due to Descartes' influence. For example, Regius' refutation of the traditional explanation of the temperament as a mixture of elements into a new homogeneous substance is closer to Gorlaeus than to Descartes, although his alternative definition is compatible with both authors:

> Therefore we define a good temperament as follows: the location, figure, quantity and motion or quiet of the insensible parts which constitute the sensible parts in such a way that it fits the actions that have to be performed. From this temperament, or from the first qualities from which it is made up, all other qualities of the human body as well as all other homogeneous or heterogeneous bodies derive their origin.[54]

As Regius then turns to explaining the various qualities of bodies within this framework, we can often point to both Descartes and Gorlaeus as potential sources. Regius' notion that heat "is the various agitation of insensible particles; cold by contrast their quiet," had been anticipated in Gorlaeus' statement that heat was not only due to the sun but was also "produced by the motion and friction of bigger particles."[55] But whereas Gorlaeus hastened to add that "how this [transformation of motion into heat] happens, escapes me, and I marvel at it, just as I marvel at many other things," and in contrast to Descartes, who in his *Principia* of 1644 was to provide a woodcut illustration to illustrate the agitation of corpuscles by the pressure of a sun rays, Regius remained silent about the mechanics of it all, allowing Schoock to express his hilarity at the vacuity of his explanations.[56]

In the case of Regius' first set of disputations, then, one is justified in wondering whether his corpuscular doctrines should be explained as a consequence of his acquaintance with Descartes, and could not be due to a previous commitment to atomistic notions.[57]

While for most of these views, the question of their debt to Gorlaeus remains unanswerable, this is not the case for those notorious disputations of December 1641, in which Regius proposed that "man was an accidental being." As stated before, in this second set of disputations, *De illustribus aliquot questionibus*, Regius entered with full force into a metaphysical twilight zone. From various reports, it would seem that these disputations were staged by Regius as a kind of showdown between the old and the new philosophy, with the public – students of medicine, philosophy and theology – forming opposing camps and shouting each other down.[58] The disputations have a powerfully Cartesian tone to them. The first disputation attacks the notion of substantial forms: there is only one such form, namely the mind; all others do not amount to more than the behavior of particles.[59] In a manner that is reminiscent of Gorlaeus, he subsequently tries to explain away the principal qualities in terms of the shapes and particles of an extended substance (which he does not here equate with a Cartesian *res extensa*). He then turns to Harvey's model of the circulation of blood, which he defends, and finishes with a praise of "mechanical" explanations.

The second disputation opens with a definition of nature from the point of view of the physician. It quickly turns once more into attack on those who "hallucinate" by defending Aristotle's view of nature.[60] In an even more clearly Cartesian manner, matter is now equated with a corporeal substance, whose essence is extension, as "matter does in reality not differ from magnitude."[61] Like Descartes, Regius also deduces from this identification of matter with extension that the small particles on which his physics rely "are not atoms, but are ever divisible" and capable of changing their shape.[62] Why and how they change their shape, and how this leads to the generation of the 'forms', is left unexplained. There is, in a Cartesian manner, a rather hand-waving gesture towards local motion as the only cause of change, but the exact mechanism of this cause must have been explained orally, during the disputation. Most startling is the way in which Regius pretends that the truth of such mechanistic explanations is demonstrated by the generation of worms in the putrefaction of cheese.[63] Spontaneous generation, one would have supposed, should for Regius' contemporaries have constituted an obvious counter-argument to a mechanical explanation of life, not its proof. However, following the doctrines of "the author of the French dioptrics," Descartes, Regius continues unperturbed, explaining that God had endowed the parts with different properties and that this explained the laws of nature.[64]

It was however the third disputation, held on 8 December 1641, that led to the Utrecht Crisis. It may seem an historical irony that the respondent, Henricus van Loon, dedicated it to the three professors of theology Gijsbert Voetius, Meinard Schotanus and Carl Dematius, as well as to Henricus Regius and the above-mentioned Jacob Ravensperger. It has been argued that this dedication was "not without provocation";[65] but as such dedications usually required prior permission, one might, to the contrary, conclude that it was not the printed text as such that caused scandal, but the manner in which the disputation evolved.[66]

In fact, from the *Testimonium Academiae Ultrajectinae*, we may gather that this third disputation was a particularly noisy affair, accompanied by a tumultuous response from an audience enriched by numerous spectators keen to witness the spectacle. It opened at once with the sketch of a Cartesian cosmos: since there is no vacuum, all motions must be circular. There is, moreover, no difference between natural and enforced motion. Finally, all motion "takes place according to laws of nature."[67]

Having thereby finished his account of matter, Regius turns to the other principle of hylemorphism, namely form. Form is in reality nothing else than the composition of particles. There is thus nothing additional to it, no "eduction of substantial forms from the potency of matter," as "those who ignore the true forms" believe.[68] But what, then, of the soul? "The special form [of man] is the human mind (*mens*)," Regius explains,

> because thanks to it, together with the general form in the corporeal matter man is what he is. 'Mind' can in no way refer to the general or material form, given that it is (as an incorporeal substance) no body, nor can come about through the motion or quiet, magnitude, location or figure of the parts.[69]

Then comes the notorious ninth thesis: "The mind and the body do not give rise to a single *ens per se*, but *per accidens*, because both are perfect or complete substances." The tenth thesis makes these same point from a different perspective, as if to persuade those who had not yet understood: "If they are called 'incomplete', one has to understand this term only from the point of view of the composite, which is brought about by their union."[70]

Having redefined man, Regius turns his attention to cosmology (defending heliocentrism) and to a theory of the elements (abolishing the traditional four-element theory), concluding with the renewed accusation of hallucination addressed to Aristotelians who maintained the traditional quaternary scheme of qualities and elements.

In short, then, in this second set of disputations, Regius combined a dualistic view à la Descartes with a great portion of Gorlaeus-style nominalism and with the tenet, also taken from Gorlaeus, that man is merely an *ens per accidens*. The scandal was perfect; and not just the university authorities were displeased, but Descartes too: "You could not have put anything harder there," he wrote to Regius.[71]

At this stage, the university authorities intervened, led by the rector, the theologian Voetius. Ever since Descartes had been praised as the century's new hero at Reneri's funeral in 1639, Voetius had been worrying about Descartes' growing influence at his university.[72] Ironically enough, he had even tried, in 1640, to persuade Marin Mersenne to refute Descartes' pernicious views, unaware of the close friendship that linked the two men.[73] The alliance between Regius, whose heterodoxy had in Voetius' eyes been demonstrated during the Naarden episode in the 1630s, and the philosophically deviant Descartes seemed dangerous, and the fact that Gorlaeus was invoked as an authority in this context made things even worse. The denial of the substantial unity of body and soul, which is not only what Gorlaeus' thesis entailed, but also what Regius seemed to propagate in his disputation and what Descartes' mind-body dualism implied, was unacceptable for a whole series of reasons, including that of the resurrection of bodies: for if the soul was a separate entity altogether, there was no need for the body, a mere heap of atoms, to be resurrected. For Voetius, in Verbeek's words,

> it was the unity of faith, and hence the moral unity of the Academy, indeed of the entire country, which was called into question: by denying, even indirectly, the dogma of the resurrection of bodies, Regius showed his overt sympathy with the Remonstrants – or worse![74]

Descartes protested that Regius was not the author of the disputed thesis about man being an *ens per accidens*, arguing that it had been inserted by some imprudent student. Regius, however, at once admitted his authorship but initially tried to defend himself by saying that this doctrine was neither his own nor Descartes' invention, but that he had taken it from Gorlaeus.[75] As we have just seen, this made things only worse. An aggressive set of corollaries crowning student disputations defended by Voetius' students on 18 and 24 December, and thus only a few days after Regius' disputations, combated Gorlaeus together with Regius and Descartes. The first draft of the corollaries, which is reproduced in the *Narratio* providing the history of the Utrecht Crisis, opens by giving the rich pedigree of the *ens per accidens* theory that we have cited earlier:

> The paradoxical claim [about man being an *ens per accidens*] made [...] by Taurellus (who was called an atheist physician by the Heidelberg theologians in their judgement on Vorstius' *De Deo*, which they sent to the delegates of the Synod of Holland in 1610), and which, due to the imprudence of youth, our compatriot David Gorlaeus took up in his *Exercitationes philosophicae*, a book he wrote in a moment when, beginning his theological studies or rather preparing himself for them, he was attacked by doubts and hesitations, is contrary not just to physical truth (which we leave to the physicists to explain), but also to metaphysics, pneumatology and theology. We therefore wish to admonish our students that given one absurdity, many others follow, and what begins as a small error, eventually grows into a large one.[76]

This first corollary thus elucidates the prehistory of Regius' thesis and points to its heretical implications, which are considered threatening particularly to theology and metaphysics. As we have examined in great detail in our earlier chapters, it throws much light on Gorlaeus and the circumstances of the composition of his *Exercitationes*.

The second corollary, which we need not analyze here, attacks Regius' views regarding the daily and annual motion of the Earth. In the third corollary, Voetius returns to Gorlaeus' atomism:

> A philosophy that denies the substantial forms of things as well as their proper and specific faculties and active qualities, and therefore the specific and distinc-

> tive natures of things (which in our days Taurellus, Gorlaeus and Basson have attempted to do), has evidently not been capable of being reconciled with the sacred physics of Moses. For this reason, we refer our students in the first place to Danaeus, Zanchi, to the authors of commentaries on Genesis, etc., as well as to the scholastic commentators of Petrus Lombardus and Thomas Aquinas.[77]

Importantly, this corollary ascribes the denial of substantial forms and their replacement by atoms or corpuscles not to Descartes, but to an earlier tradition that we have investigated above. By the same token, Regius' own unwillingness to yield to Descartes' pressure to give up his view of man as an *ens per accidens* even in the turbulences that followed, demonstrated his rootedness in a pre-Cartesian ontology and his fidelity to Gorlaeus, "his first source of inspiration."[78]

To the 'Remonstrant' axis of Gorlaeus and Taurellus, Voetius added Sébastien Basson (c. 1573-after 1625), a Calvinist physician and university teacher, whose book was much read in the Netherlands and, in 1649, saw a second, Amsterdam edition. Isaac Beeckman was not alone in having appreciated a number of ideas that he encountered in Basson's work (while rejecting others and lamenting the author's lack of mathematical understanding), but Constantijn Huygens had personally recommended Basson's atomist natural philosophy to Descartes, who in 1629 wrote to Marin Mersenne: "As far as rarefaction is concerned, I agree with this physician [Basson], and now side with him concerning the foundations of philosophy; but maybe I do not explain the ether like him."[79] Given that Gorlaeus had influenced Regius, and Basson had influenced both Descartes and Regius, they were, in Voetius' eyes, all members of the same heterodox gang.[80]

In other words, then, in Vorstius' eyes, the doctrines proposed by Regius and Descartes belonged to a current that had reached the Netherlands earlier. But does this observation not allow us to go beyond the ascertained influence of Basson on Descartes and of Gorlaeus on Regius, and to take the additional step of stipulating an influence of Gorlaeus on Descartes? Admittedly, Descartes himself protested, in his Letter to Dinet of 1642, that he had never heard of this author.[81] Also, as we have seen, Gorlaeus' atomism is not a direct ancestor to a mechanical conception of material interactions, as he attributes little importance to atomic shapes and motions and allows for travelling 'real accidents'. Still, his rejection of substantial forms, his insistence on the agency of invisible corpuscles, his radical separation of mental entities from physical atoms, his understanding of substances and their modes and – even more importantly for Descartes – his attempt to derive his physics from an ontology, all render the assumption that Descartes would have found many of his ideas congenial quite attractive.[82] Having inscribed himself at the same universities as Gorlaeus – Franeker first, then Leiden – and having lived with or frequented persons who possessed and used copies of the *Exercitationes*, it would certainly not be far-fetched to imagine that Descartes had come across Gorlaeus' name or even read his work. Moreover, as Hattab has argued, in Descartes' eyes, "Gorlaeus' thoroughgoing and metaphysically grounded atomism" may have "provided the conceptual resources for eliminating any vestiges of the Aristotelian hylemorphism" and at the same time allowed him to connect "the Gorlaean theory of modes to his [own] dualism" as the "metaphysical foundations" of his physics.[83]

Might Voetius, then, have been right to see a shared ideology behind the thesis that triggered the Utrecht Crisis? And was it wrong of Regius to understand Descartes' substantial dualism as resulting in the view that man was an accidental aggregate of body and soul? And what was Descartes' own view on the matter? As Robert Pasnau points out:

> To be sure, [Descartes] does not want to be read as defending Platonic dualism along Gorlaeus' line. But is this because he does not believe it to be true, or because he does not dare say it, even if he thinks it?[84]

Irrespective of the existence of such a direct influence, it is obvious that contemporaries perceived a link between Gorlaeus, Regius and Descartes. On 24 December 1641, both Gorlaeus and Basson were being refuted in a disputation conducted under Voetius, which defended the existence of substantial forms. Voetius there directly links their rejection of such forms to the idea, now also shared by Descartes and Regius, that animal bodies operate liked clockworks.[85] In the same disputation, Taurellus and Gorlaeus are rejected for their notion that man is a composite being, with philosophical as well as theological arguments.[86] In the same spirit, in his polemical anti-Cartesian tract of 1643, the *Admiranda methodus*, Martin Schoock confirmed that Gorlaeus and Taurellus were authors with whom Dutch students and teachers were generally acquainted.[87] This did not mean, however, that Schoock liked either of them. In fact, he derided Descartes' seemingly arrogant attempt to replace Aristotle by comparing it with earlier attempts made by "Basson, incompetent falsifier of the views of ancient philosophers; Taurellus, whose maniacal debating lust has edged up to the limits of atheism, and Gorlaeus, his student who followed him with uneven steps."[88] The same association occurs when Schoock writes: "And let me not say anything of all those academies and schools in Europe which all, I am sure, will reject the philosophy of Descartes and prohibit it at the same time with the delirious thoughts of Taurellus, Gorlaeus and Basson."[89]

The evidence of the Utrecht Crisis allows us to conclude that in the eyes of Voetius and his associates, around 1610, a subterranean link had come into existence between Vorstius' theology with its physicalist claims concerning God, on the one hand, and an atomistic ontology, on the other. Even in the years before the outbreak of the Utrecht Crisis, Voetius had held university disputations combating 'atheism'; in these disputations, Arminians and Socinians in general, and Taurellus and Vorstius in particular, were regularly attacked.[90] Voetius remained so aware, or suspicious, of this alliance that he could not help but conceive Descartes' corpuscular explanations as an attempt to bring Arminian or Socinian ideas back into circulation. That the theologically heterodox Regius, friend of Descartes, relied in one of his disputations on Gorlaeus' Taurellian thesis of man as an *ens per accidens* must obviously have confirmed Voetius' fear.[91]

4.3. GORLAEUS FORGOTTEN AND REDISCOVERED: A CONCLUSION

In the period 1641-43, Gorlaeus' name and some of his philosophical views were thus once more discussed and debated. Despite the fact that the author of the *Exercitationes* had died thirty years earlier, the Dutch academics involved in the controversy either knew or intuited how to place these views ideologically, and some, like Voetius, even remembered the author and his religious affiliation personally. As his academic disputations of the 1630s document, for Voetius, the Arminian question was still a burning issue, and seeing positions he associated with the Remonstrant movement being defended at his own university, Utrecht, was intolerable in his eyes. This is why Regius' explanation that he had taken the thesis concerning man as an accidental being from Gorlaeus' treatise was, as far as Voetius was concerned, not a mitigating, but instead an aggravating circumstance.

Merely ten years later, in 1651, Gorlaeus' other treatise, the refreshingly short *Idea physicae*, was published in Utrecht. But alas, judging by the extant copies and the contemporary references to it, almost no one bought or read this book. Admittedly, its print run must have been quite small, but the silence with which Gorlaeus' treatise was greeted also had intellectual reasons. Not only did the *Idea physicae* contain no essentially new ideas with respect to the *Exercitationes* but, more importantly, by 1651, Gorlaeus' philosophy began to look somewhat obsolete. As for his atomism, there were newer and more exciting works around. Whoever wished to have metaphysical arguments for atomism could now turn to Pierre Gassendi, whose *Syntagma philosophiae Epicuri* had appeared in 1649. In the same year, Ludovicus (III) Elzevier republished Gassendi's *Exercitationes paradoxicae* (1624), just as he had republished Sébastien Basson's atomistic *Philosophia naturalis* (1621) the year before. Who instead preferred a metaphysical system leading to a corpuscular philosophy (which for all physical purposes was equivalent to atomism), could rely on René Descartes' *Principia philosophiae* of 1644. Who, in turn, wished for a corpuscular theory of matter that dispensed with a metaphysical pedigree, could buy the *Fundamenta physicae* (1646) of Descartes' erstwhile friend Henricus Regius. Who desired to combine an allegedly revived Democritean system of atomism with a chemical theory, could use Jean-Chrysostôme Magnen's *Democritus reviviscens*, which had first appeared in 1644 in a handsome quarto edition in Pavia, but was more cheaply reprinted at Leiden in 1648. Finally, Sennert's *Hypomnemata* of 1636, which represented the culmination of this author's conversion to an atomistic model of matter, was reprinted in 1650 both in Frankfurt and Lyon. In other words, the years 1644 to 1650 witnessed a veritable explosion of physical theories that relied on the shape and motion of microscopically small particles, whether these were defined as indivisible atoms or theoretically divisible corpuscles. Gorlaeus' work had prepared the grounds for some of these systems – but having done this preparatory work, so it must have seemed to most mid-century readers, one could now dispense with him.

Moreover, Gorlaeus' theory of matter, which started from a definition of *ens* that applied to God as much as to material particles and which featured mysterious 'real accidents' travelling from atom to atom, may have begun to look unconvincing in an age that had come to embrace the hope that geometrically defined, quantifiable and therefore visualizable particles would allow for a mathematization of nature and its laws and for a deduction of secondary, sensory qualities from the primary, geometrical properties of the ultimate corpuscles. In this sense, one could claim that it was the legacy of Isaac Beeckman, not of David Gorlaeus, that had come to triumph in the woodcuts that adorned the physics books of Descartes and a number of his followers.

Indeed, as Klaas van Berkel has repeatedly emphasized, Beeckman's thinking has strongly visual overtones – a visuality ('aanschouwelijkheid') that would become even more strongly emphasized in the geometrical physics of Descartes, whom Beeckman had first met in November 1618.[92] In fact, the profound differences that separate Gorlaeus' from Beeckman's atomism are both perplexing and startling, given the close parallels in the lives of the two Dutchmen. Beeckman (1588-1637) was merely three years older than Gorlaeus, and like him also descended from a Brabant family that had migrated to the Netherlands. In 1607, four years before Gorlaeus, he enrolled at Leiden's faculty of theology, but ended up doing more mathematics than anything else and after a mental breakdown had to return home. In the fall of 1609, he moved once more to Leiden to continue his theological studies, but – possibly because of the intensely unpleasant atmosphere that reigned at the faculty at the culmination of the Vorstius crisis – he left Leiden again in the summer of 1610, not long before Gorlaeus arrived in town. Indeed, as Van Berkel writes,

> it is curious to realize that while Isaac Beeckman left Leiden to return to Middelburg and [his brother] Jacob Beeckman travelled from Leiden to Franeker, the young David Gorlaeus moved from Franeker to Leiden, where he would write his *Exercitationes* and his *Idea*, both tracts in which atomistic doctrines are used [...]. Beeckman later owned a copy of the *Exercitationes*.[93]

There are in fact a few elements that the two young, theologically trained Dutch atomists share, for example their dislike of Aristotle's natural philosophy and their use of a nominalist phraseology in their search for a limitation of principles. Like Gorlaeus, Beeckman reproaches scholastic philosophy for its unnecessary proliferation of forms, pleading instead for an economy of principles ("Male fit per plura quod bene fit per pauciora").[94] And like Gorlaeus, he inclines – at least during a certain period of his life – towards an atomistic model that allows for a dramatic reduction of material principles. But for the rest, there is very little the two thinkers have in common. Steeped in his experience as an artisan and later also as a physician, Beeckman is interested in explaining concrete phenomena; for his figured atoms, he draws his inspiration from Democritus and Lucretius, rather than from Protestant ontologists; and, above all, he attempts to combine his theory of matter with mathematics.[95]

"There are very few physico-mathematicians," Beeckman sighed after having encountered Descartes in 1619.[96] 'Physico-mathematics' is in fact the key word for understanding Beeckman's and Descartes' shared project. It was only later that Descartes felt the need to have his physico-mathematical natural philosophy grow out of the metaphysical roots of a first philosophy.[97] What Descartes may thus owe to Gorlaeus or to those Dutch friends who were acquainted with his *Exercitationes*, is the idea that one's physical principles had to be deduced from a first philosophy, or ontology, and not induced from experience. But maybe Descartes' debt is much bigger and more specific than this. For example, Helen Hattab has recently argued that Descartes embraced "Gorlaeus' substance/mode ontology," by explaining the entire natural world in terms of a single corporeal substance and its *modi*.[98]

However that may be, the genesis and genealogy of Descartes' ideas was of little interest to his followers. They were not curious about the evolution of their idol's thought (about which Descartes himself was notoriously solipsistic), but, if anything, they took an interest in the vicissitudes of his life and the adversities he encountered. As Gorlaeus' name had come up in one of the most intensely controversial moments in Descartes' intellectual life, he was henceforth often reduced to this marginal part he had played in the early history of Cartesianism. It is only in this respect, for example, that we encounter him in Pierre Bayle's *Dictionnaire*. In the passage that is dedicated to him, he is presented as an otherwise unknown philosopher whose views had however managed to trigger the Utrecht Crisis. Although this was by itself sufficient for Bayle to feel some sympathy for this early opponent of the stubborn Voetius, he did not hide the fact that he had no patience for Gorlaeus' philosophical language:

> *Ens per se, Ens per accidens*, these are inexplicable expressions, veritable jargon of the Spanish logicians, which does not mean anything.[99]

It is obvious that by the end of the seventeenth century, Gorlaeus' intellectual world had in most quarters all but vanished, as had that of Voetius.[100] Of course, Northern Europe's *salon* avant-garde and the level of discourse that dominated the epistolary exchanges of the Republic of Letters must not be mistaken for a philosophical consensus. The philosophy that was taught and learned depended strongly on local traditions and circumstances. In chapter 1, we encountered the Franeker professor of philosophy Arnold Verhel who, in 1662, was still inveighing against "the zeal of the Ramists, the gainsaying of the Gorlaeans, the high-browed arrogance of the Cartesians, and the authority of certain teaching doctors."[101] While in many places, it was above all Descartes, Hobbes and soon also Spinoza that one combated, this does not mean that there could not be institutions at which Ramus and Gorlaeus were still authors that were controversially debated.

Verhel's anti-Gorlaeanism is, however, a bit of a surprise. After all, Gorlaeus' terminology, which around 1610 or 1620 was cutting edge thanks to its original combination of a German ontological terminology with sixteenth-century Italian natural philosophical views, had for most readers lost its allure of novelty by the 1650s. The philosophical avant-garde, heavily influenced by Descartes, now tended to approach natural philosophy from an epistemological point of view. While Descartes had introduced his own ontology by way of the epistemological procedure first presented to the public in the *Meditations*, Gorlaeus had started with an ontology and had only subsequently added his nominalist epistemology. For Cartesians, at any rate, Gorlaeus' ontological principles must have seemed unwarranted, and its vocabulary scholastic gibberish. As Gabriel Daniel, in his *Voyage du monde de Descartes* (1691) was to put it, such old and antagonistic currents of thought as Thomists, Scotists or Nominalists were now only debated at some backward teaching institutions. Modern thinkers, by contrast, had lost interest in these old distinctions:

> We put them [the earlier philosophical currents] all in the same category, and in the same party, which we call 'ancient philosophy', to which we contrast 'the philosophy of Descartes', or 'the new philosophy'. You [Descartes] have had the good fortune to erase, in a way, everything that appeared from the new philosophers at the same time as you.

Just as it had been customary in sixteenth-century Spain to call all heretics 'Lutherans', whatever their particular sect, so Daniel adds, one now called 'Cartesians' all those who tried to elaborate a new physics, including Gassendi (who in many respects could have been regarded as Descartes' rival).[102]

These various circumstances help us explain why, until his rediscovery by Kurd Lasswitz, Gorlaeus fell into oblivion; if he was remembered at all, it was as the transmitter of Taurellus' obscure definition of man as an accidental being to the Cartesian disputes of the mid-seventeenth century, and thus as a bizarre figure who had featured during the birth pangs of Cartesianism.[103] He thereby become a figure in what the historiography of early modern thought came to define as 'a period of transition' – a transition away from a medieval and Renaissance world dominated by scholastics, in which the sixteenth- and early-seventeenth-century *novatores* fulfilled the role of an avant-garde that paved the way for the new systems of a Descartes, Hobbes, Leibniz, Spinoza or Newton. Stephen Menn writes about these *novatores*:

> [They] produced such new philosophies because there was a demand for a new philosophy, that is, a current expectation of what a philosophy should do, and a sentiment that the old philosophy was not doing it properly. Indeed, one may say that the chief philosophical legacy which the sixteenth century bequeathed to the seventeenth was not any particular new philosophy but this *expectation* of a new philosophy.[104]

As we have seen in chapter 1, Gorlaeus was routinely attributed an honorable place in the gallery of these *novatores*, and sometimes even viewed as a precursor of Cartesianism. Morhof's claim that Gorlaeus "had recognized before Descartes what Descartes later wanted to make appear as his own doctrines" is obviously wildly exaggerated.[105] Reimmann's alternative claim that "Cartesians afterwards accepted most of Gorlaeus' theses into their system" is, for chronological reasons, equally implausible.[106] On the basis of the evidence that we have presented in this book, I think it would be fair to combine Morhof's and Reimmann's assessments into the following, more modest claim: Gorlaeus' derivation of physics from metaphysics, together with his distinction between indivisible material and spiritual entities, helped to prepare the ground in which Descartes' system, as presented in the *Principles of Philosophy*, could take root and in which his mind-body dualism could develop. A direct influence of Gorlaeus' philosophy on the development of Descartes' thought is difficult to prove, although it is for the above-mentioned reasons quite plausible. By contrast, Gorlaeus' influence on Descartes' first Dutch friends can readily be documented. To this extent, then, we may speak of Gorlaeus as a precursor of Descartes' "metaphysical physics" (to cite Dan Garber's fortunate expression) and possibly even of Descartes' specific view on substances and modes (as Helen Hattab has argued).

In this sense, the findings of this book call out for some adjustments to our available histories of metaphysics, its relation to physics and the prehistory of Descartes' philosophy. For they contradict the assumption that seventeenth-century Dutch attempts to include a definition of God in a reformed metaphysics are due to the *Disputationes metaphysicae* (1597) of the Spanish Jesuit Suárez.[107] Taurellus' *Philosophiae triumphus* not only preceded Suárez' *Disputationes* by more than 20 years but, as Leinsle has shown in his masterful history of Protestant metaphysics, this philosopher "established on a nominalist basis a metaphysics that was both unique and independent" of Spanish metaphysics.[108] Given that Taurellus' impact on Gorlaeus' own work preceded Suárez' introduction into the Dutch syllabi, the currently available histories of seventeenth-century Dutch metaphysics will in this respect require some modification.[109]

As for that other question relating to Gorlaeus' role in the history of atomism, it is far more complex. He is, of course, one of the earliest neo-atomists and as such deserves the historical place that has been granted to him since his rediscovery by Kurd Lasswitz. However, given the complexity of the phenomenon of the early modern revival of atomism, it is not evident exactly how one must define his role.

This difficulty has to do with the problem of early modern atomism as a whole. Simply put, among the key elements that separate an Aristotelian or scholastic understanding of nature from that of modern science, our history books routinely single out theories of matter for their importance. As has been explained in chapter 2, the difference between the two views of nature may be sketched as follows. According to Aristotelian hylemorphism, natural substances are in the last analysis understood as composites of prime matter and substantial forms, where the latter inhere in the former only transitorily. Although the hylemorphic and the atomic understanding of matter would therefore appear to be diametrically opposed to one another, it would nevertheless be misleading to assume that there was a precise moment in the history of early modern science when a paradigmatic shift from the first model to the second occurred. Thomas Kuhn once described his personal experience of how, "one memorable" and "very hot" "summer day," he managed to break into the logic of Aristotelian physics, interpreting this experience as the inverse of the *gestalt* switch taking place in the Scientific Revolution.[110] While this personal experience is fully credible, it does not capture the spirit of the multiple and complex transformations that characterized the evolution of physics in the early modern period. There are some precise arguments for why this is not so.

The first argument against the assumption of such a radical rupture is that the atomic theory never entirely replaced hylemorphism, some version of which survived in chemistry and hence in natural philosophy until the end of the nineteenth century. As late as 1875, for example, the chemist Thomas Sterry Hunt protested that mixture was no "juxtaposition, as conceived by the atomistic chemists," but has to involve "interpenetration," as the philosophers Aristotle, in antiquity, and Hegel, in the early nineteenth century, had rightly stressed.[111]

The second argument has to do with the fact that in local contexts, beginning in fifteenth-century Italy, there existed some currents within Aristotelianism itself which took chemical mixtures to possess a corpuscular structure and which therefore combined atomic with hylemorphic notions.[112]

But not only did some local forms of atomistic or corpuscular thinking precede the seventeenth-century downfall of Aristotelianism by a long stretch; and not only did atomism remain a contested hypothesis until the end of the nineteenth century; the third argument to keep in mind is that early modern atomic and corpuscular ideas constituted a phenomenon of such heterogeneity that it would be quite implausible to consider it a single, unified paradigm. Giordano Bruno's ensouled monads, René Descartes' indefinitely divisible particles of *res extensa*, Pierre Gassendi's indivisible atoms with their hooks and eyes, and the chemical atoms and corpuscles that were proposed in the period between Daniel Sennert and Robert Boyle have little in common with one another. The best proof of this fundamental heterogeneity is found in the way in which Voetius, when combating Regius', Descartes' and Gorlaeus' replacement of substantial forms with atoms, invoked Sennert's atomism as an acceptable alternative, according to which the atoms were the very carriers of the substantial forms![113]

Although atoms are nowadays observed, analyzed, combined, and even split in laboratories and nuclear power plants, the concept 'atom' was not developed in what we would nowadays recognize as a 'scientific' context. In fact, the reality of atoms remained a bone of contention between groups of philosophers, chemists and physicists until the very moment that atoms were first proven to exist experimentally – a moment that roughly coincided with the experimental proof that the atom was not an atom at all, as it was not indivisible (*a-tomos*), but possessed a composite structure.

It may therefore be held that when the ancient Greek concept of 'atom' was revived in the Renaissance and the early modern period by chemists, physicians, natural philosophers and theologians (we may now add) who wished to explain with it certain natural phenomena or define the relation of God to the World, there existed as many valid empirical and logical arguments against the existence of indivisible chunks of matter as there existed arguments in its favor.[114] From this, it necessarily follows that early modern proponents of atomism tended to have reasons that went beyond recognizably scientific ones to promote the existence of atoms. There were authors who, dissatisfied with the Aristotelian notion of purely potential prime matter or the unclear provenance of substantial forms, sought in rivalling ancient schools of thought for an alternative. There were those others who believed, more specifically, that Democritus, Hippocrates and Plato had developed a medico-philosophical theory of matter that was more ancient, venerable and therefore superior to Aristotle's. Some even believed that atomism went back to Moses himself, and that Aristotle's anti-atomism represented a corruption of an original, divine philosophy. There were yet others who favored atomism because they liked Democritus (whom they took to have been one of the fathers of chemistry), or Epicurus (whose ethics they preferred to Aristotle's), or Lucretius (the style of whose didactic poem, *De rerum natura*, they admired and imitated). There were also those who had religious reasons for preferring atomism: like the members of the medieval Islamic school of the Mutâzilī in Basra, they felt that in a world composed solely of God and atoms, all causality could be attributed exclusively to God, and that this was more attractive than having to negotiate the complex relationship of the primary (divine) causality with the secondary (natural) causality that characterized the Christian-Aristotelian synthesis.

Where, then, must we place Gorlaeus in this intricate and fairly elusive story? We recall from chapter 1 Kurd Lasswitz' bafflement at reading the *Exercitationes*, which in his eyes offered an unusual, indeed unique, type of justification for the existence of atoms. Historiographical claims to the superiority of Democritus are as absent from it as are experimental, chemical arguments or the kind of reasoning from a divine geometry that one finds in Giordano Bruno's Cusanian arguments from the coincidence of opposites. What is indeed unique about Gorlaeus is his universal atomism, which included God, angels, souls and material indivisibles as its four types of indivisibles. Sure enough, Bruno's atomism is also metaphysical, and in light of the sojourn of Gorlaeus' teacher De Veno in the same prison to which Bruno was confined, it is tempting to search for Brunian echoes in Gorlaeus' works. However, there is nothing in the *Exercitationes* that resembles Bruno's archetypal understanding of the atom and God as two extremes mirroring each other; nor are Gorlaeus' atoms ensouled, dynamically unfolding entities. For some years in the late 1580s, Taurellus and Bruno were both teaching at German universities, and both nurtured atomistic conceptions that possessed strong theological conceptions. But Bruno's immanent deity that grew, as it were, out of the ensouled atom is a far cry from Taurellus' transcendent God and his material atoms. Gorlaeus is clearly the heir of the latter.[115]

It has been argued in this book that Gorlaeus wrote his *Exercitationes* under the influence of Taurellus, as an attempt to establish an ontological basis from which metaphysical and theological issues could be resolved in the hope that it would benefit the Arminian conception of the relation and interaction between God and his Creation. Still, we have seen that neither his *Exercitationes*, nor certainly his *Idea physicae*, directly address theological matters, but at best allude to them. As a new student of theology, Gorlaeus had to confine himself to metaphysics and natural philosophy or physics. However, by necessity, natural philosophy implicitly or explicitly relied on concepts taken from theology (as the science of God, Creator of all natural beings) and metaphysics, notably ontology (as the metaphysical doctrine of 'being' in general). And conversely – as the harsh reaction of Voetius demonstrates – changes in the principles of natural philosophy could not but have severe repercussions in theological doctrine. In a period that saw the confessionalization of physics – physics serving specific confessional concerns and theological doctrines driving physical doctrines in specific directions – the denial of substantial forms and the mechanization of causality with its abolition of teleological explanations could be, and was, in many quarters perceived as a "crisis of causality," as Han van Ruler has documented in convincing detail.[116]

Given the extraordinary precocity and philosophical originality that Gorlaeus displayed at age 20, it is tempting to imagine him, later in life, producing a fully fledged physico-theological alternative to the model that was at the time being taught at the universities. As it stands, we possess an impressive testimony to an intellectually bold, independent and versatile young man, who died far too early; but yet managed to contribute, albeit posthumously, to the evolution of metaphysics and physics between 1620 and the mid-century, and to whom time was not given to work out a mature system that could compete with those that other members of his generation, such as Descartes, Gassendi or Hobbes, worked out in the 1640s, at an age that tends to be more suitable for the composition of systematic masterpieces.

Notes

CHAPTER I

1 "Anno 1612 den 27 April is gestorven den Welgeleerden en Seer Verstandigen Ionghelinck David van Goorle alhier begraven."

2 "Honori & memoriae ornatissimi Juvenis Davidis Gorlaei:
Heic est sepultus ille flos juventutis / Gorlaeus, ipso in vere raptus aetatis, / Ad summa surgens orsa laudis antiquae. / Mors quae nihil sublime fert diurnare / Terris ademit debitum decus coelo, / Nam terra tanti non erat capax doni. / Praelustris illa mens et aura coelestis / Exuta vinclis atque corporis mole / Coelum petivit, unde duxerat semen. / Ibique natum patre cernit aeterno, / Qui morte crimen expiavit humanum, / Mundi sequester Christus et dator pacis. / Haec est salus, hic apex boni summi, / Quo nostra mens vovere nil potest maius." Coats of arms were in that region systematically removed by the order of church authorities in 1796. On the possible link between the text of this poem and Gorlaeus' understanding of the relation of body and spirit (or soul), see below, pp. 39-40.

3 Gorlaeus, *Exercitationes philosophicae, quibus universa fere discutitur philosophia theoretica, et plurima ac preacipua Peripateticorum dogmata evertuntur* (Leiden, 1620).

4 Gorlaeus, *Idea Physicae, cui adjuncta est Epistola cuiusdam Anonymi de Terræ Motu* (Utrecht, 1651).

5 Andrea, *Bibliotheca Belgica*, 173.

6 Foppens, *Bibliotheca Belgica*, 229.

7 Dijksterhuis, *Mechanisation*, 4: 227.

8 Mersenne, *La vérité des sciences*, 109-10: "Aristote est un Aigle en Philosophie, les autres ne font que comme des poussins, qui veulent voller avant que d'avoir des ailes."

9 Mersenne, *Impiété des Déistes*, 1: 237-38: "… en faveur de toutes les véritez contre toutes sortes de mensonges, dedans laquelle j'examineray plus diligemment ce qu'ont advancé Gorlée, Charpentier, Basso, Hill, Campanelle, Brun, Vanin et quelques autres." Gorlaeus is once again listed in Mersenne, *Quaestiones celeberrimae*, col. 1838.

10 *Testimonium Academiae Ultrajectinae*, "Corollaria," Corollarium 1: "Assertio Taurelli … quam imprudentia juvenili ex illo adoptare voluit popularis noster David Gorlaeus … ." For the context of this "corollary," see Verbeek, *Querelle*, "Introduction," and Bos, *Correspondence between Descartes and Regius*. This episode will be analyzed in detail below, pp. 148-53.

11 Verhel, *Speculum philosophiae primae entis*, dedication: "Nescio quae fatalis huius miserandi aevi calamitas studia haec Philosophica invasit […]. Insuper Aristotelem ipsum, Primae Philosophiae parentem et Philosophorum Principem, conviciis et dicteriis proscindunt. Infremuit adversus eum Ramaeorum zelus, Gorlaeorum contradicentia, Cartesianorum supercilium, docentium quorundam Doctorum auctoritas." Quoted from Vriemoet, *Athenarum Frisiacarum libri II*, 201. See also Galama, *Wijsgerig onderwijs*, 56-61.

12 Morhof, *Polyhistor*, 2: 71: "Exiguus liber est, sed tamen cum aliquo ingenio scriptus. Incipit a Philosophia in genere, hincque ad Metaphysica & Logica pergens, multa illic statuit a Peripateticis

discrepantia. Procedit deinde ad Physica, in quibus suas sequitur hypotheses, Aristotelicasque impugnat. Quarum praecipue sunt illae: coelum nihil aliud esse, quam extensionem aeris; duo fecit mixtorum elementa, terram & aquam; ignem e numero elementorum excludit, merumque facit accidens, quod etiam primarium Cartesianorum dogmatum est." A similar characterisation is found *ibid.*, 2: 245.

13 Burmannus, *Traiectum eruditum*, 106: "Philosophus non incelebris fuit, iisque, qui Aristotelem impugnare ausi sunt, adnumeratur."

14 Garber, *Descartes' Metaphysical Physics*, 61.

15 Bayle, *Dictionnaire*, vol. 7, s.v.: "Gorlaeus (David), natif d'Utrecht, a vécu dans le XVIIe siècle. Il publia quelques livres de philosophie, où il s'écarta de l'opinion ordinaire des écoles." This biographical introduction is followed immediately by the Regius incident: "Régius, disciple de M. Descartes, se voyant harcelé pour une thèse qui concernait l'union de l'âme et du corps, allégua qu'il s'était servi de propres termes de Gorlæus. Cela ne lui servit de rien, et fut cause que Voëtius, professeur en théologie, flétrit autant qu'il lui fut possible les sentiments de Gorlæus."

16 Morhof, *Polyhistor*, 2: 273: "Laudem certe meretur, quod ante Cartesium ista viderit, quae postea Cartesius dogmata sua esse voluit."

17 Reimmann, *Versuch einer Einleitung*, 3: 563: "Denn alle diese Einwürfe [*sc.* Voetius'] haben die Cartesianer, welche die meisten Hypotheses Gorlaeanas hernachmals in ihr Systema philosophorum aufgenommen, dem Voëtio beantwortet."

18 See, e.g, Van Berkel, *Isaac Beeckman en de mechanisering*, 292-301; idem, *Mechanical Philosophy*.

19 Cudworth, *True Intellectual System of the World*, title of ch. 2: "that the *Democritick philosophy*, made up of *Corporealism* and *Atomicism* complicated together, is essentially atheistical." For a Latin usage, see Saguens, *Systema Eucharisticum*, 278: "Hinc etiam nec Arriaga, nec Poncius incusatus fuit *Atomismi*, licet uterque docuerit levitatem & gravitatem, raritatem & densitatem & alias qualitates ex istis pendendes non esse, nisi modaliter distinctas a suis substantiis, & ideo esse realiter indistinctas."

20 Pasnau, *Metaphysical Themes*, 84 and 88.

21 Hannequin, *Essai critique*, 2: "Ainsi les théories contemporaines sont sur ce point d'accord avec l'histoire, elles consacrent la préponderance, dans le domaine scientifique, de l'hypothèse atomistique."

22 On Lasswitz, see Lüthy, "Atoms and Corpuscles."

23 Lasswitz, *Geschichte der Atomistik*, 1: 333-35; 455-73; 333: "Eine Monographie über Gorlaeus und dieses wichtige Jahrzehnt wäre sehr erwünscht."

24 Land, *Geschiedenis van de wijsbegeerte*, 117: "... doch die, door nauwer te letten op natuurverschijnselen en op de beginselen der bewegingsleer, veel grooter welslagen op zijn ondernemen zou hebben bereikt dan zijn speciale opleiding voor de kerk hem schijnt vergund te hebben."

25 Sassen, *Geschiedenis van de wijsbegeerte*, 131: "... de eenzame figuur van David van Goirle, die zich reeds lang vóór den Franeker hoogleraar [Holwarda] op enkele belangrijke punten van de Aristotelische physica had losgemaakt en evenals hij getracht had die door een atomistische natuurleer te vervangen."

26 Land, *Geschiedenis van de wijsbegeerte*, 117: "Misschien vindt nog eens een bevoegd schrijver tijd om de verdiensten en tekortkomingen aan te wijzen van een die zich zeer beijverd heeft een van de hervormers der wetenschap te worden ... ."

27 Jaeger, "David van Goorle."

28 Van Melsen, *From Atomos to Atom*, 77 and *passim*.

29 Gregory, "David van Goorle."

30 Hooykaas, *Geschiedenis der natuurwetenschappen*, 133: "Isaac Beeckman en de Utrechtenaar David van Goorle (beiden stammend uit een Vlaams-calvinistisch geslacht) behoorden tot de eersten die zich bij fysische verklaringen geheel baseerden op de atoomleer."

31 Gorlaeus, *Exercitationes*, XIII.ii.247: "nihil reale esse in corporibus præter atomos, ex his omnia corpora sunt composita." Text passages in the *Exercitationes* will be referenced as follows: the upper-cap Roman numeral indicates the number of the *excercitatio*, the lower-cap Roman numeral the section numbers, and the Arabic numbers indicate the page number. "Gorlaeus, *Exercitationes*, I.iii.15" thus refers to *exercitatio* I, section iii, page 15. Where there exists only one section, the reference will be given thus: "Gorlaeus, *Exercitationes*, VI.[i].99." References to the *Idea physicae* will be given with upper-cap Roman numerals referring to chapters, paragraph signs referring to paragraphs, and Arabic numbers to the pages. "Gorlaeus, *Idea*, III.§7.19" thus refers to chapter III, paragraph 7, page 19. For the *Idea*, which is an exceptionally rare book, I have used the copy of the British Library (shelf mark 531.a.11) and Peeter's trilingual edition, *Gorlaei Idea Physicae*.

32 Jaeger, "David van Goorle," 206.

33 Partington, "Origins of the Atomic Theory," 260-61.

34 Jaeger, *Elementen en atomen*, 137-38.

35 Whyte, *Essay on Atomism*, 48-49.

36 Van Kleef, "Redactioneel," 1-2, 2: "De huidige subfaculteitsvergadering heeft zich bij het kiezen van een naam voor het chemiecomplex schuldig gemaakt aan laksheid, het a bout portant vaststellen van een naam, zonder verder iets meer van de man achter die naam te weten te willen komen, en hieruit voortvloeiende een poging tot geschiedsvalsing. Driewerf foei!"

37 Column without title in *Chimica*, 1.5 (1970), 2: "... is dit een bewijs van het feit dat studenten wel degelijk inspraak hebben bij ons in de subfaculteit."

38 Reinoud, "Gorlaeus Laboratorium," *Chimica*, 6.2 (1970), 2: "Wie Gorlaeus was of is, is mij niet bekend, maar daar zal waarschijnlijk in de naaste toekomst een onderzoek naar worden gedaan."

39 Tremerius & Fringilla, "Gorlaeus." See Van Nieuwenburg, *Korte geschiedenis*. I wish to thank Jos van den Broek for telling me about this episode and sending me the relevant documentation. On name-giving in Leiden, see Van den Broek, *Van Albinusdreef tot Zeemanlaan*.

CHAPTER 2

1 Gorlaeus, *Idea physicae*, structure: "Cap. I: Quod agit de constitutione physicae et naturae. Cap. II: De internis (vulgo ita dictis) naturae principiis. Cap. III: De externis rerum naturalium principiis. Cap. IV: De composito, quanto, continuo. Cap. V: De motu, loco, tempore. Cap. VI: De coelo. Cap. VII: De elementis et mistione. Cap. VIII: De meteoris. Cap. IX: De metallis, anima, vita et morte. Cap. X: De anima vegetativa. Cap. XI: De anima sentiente. Cap. XII: De qualitatibus praecipuis sensus afficientibus. Cap. XIII: De anima humana."

2 Gorlaeus, *Exercitationes*, structure. "Ex. I: no title [section titles: (i) Quid sit philosophia; (ii) De animae perfectione; (iii) De philosophiae distributione; (iv) De metaphysica peripatetica]. Ex. II: De ente. Ex. III: De distinctionibus. Ex. IV: De universali et singulari. Ex. V: De accidente. Ex. VI: De quantitate. Ex. VII: De qualitate. Ex. VIII: De relatis. Ex. IX: De motu. Ex X: De loco. Ex. XI: De tempore. Ex. XII: De composito. Ex. XIII: De atomis. Ex. XIV: De materia et forma. Ex. XV: De rerum ortu et interitu. Ex. XVI: De coelo. Ex. XVII: no title [sections i-v: about the elements; section vi: Globum terrestrem non moveri]. Ex. XVIII: De anima."

3 Peeters, *Davidis Gorlaei Idea*, 3, was the first scholar to spot the two references to Galileo.

4 Skinner, "Meaning and Understanding."

5 I have the reference from Simoni, ed., *Catalogue of Books*, 244. The shelfmark is 8405.de.18. I would like to thank Helen Hattab for showing me a copy of her microfilm version of this title page.

6 See *Nieuw Nederlandsch Biografisch Woordenboek*, s.v. See also Burger, "De boekverkoper Commelin." Van Huisstede & Brandhorst, *Dutch Printer's Devices*, 1: 348, locate Commelin's widow at Amsterdam.

7 On the Commelins' presence in Leiden, see Gruys, "Een Leidse veilingscatalogus." The connection with Ganne with Jacob and Isaac Commelin becomes apparent in the Testament of Jacob Commelin (see Regionaal Archief Leiden, Oud Notarieel Archief 210, nr. 174 [31 August 1624]). I wish to thank Paul Hoftijzer for drawing my attention to this source.

8 They are, for example, absent from the *Adresboek Nederlandse drukkers*, which otherwise lists, phonebook like, the addresses, functions and signs of all Dutch printers of the Golden Age. In Briels, *Zuidnederlandse boekdrukkers*, 584, Van Westerhuyzen appears, for example, as one of about twenty printers and publishers summoned in 1626 by Leiden's burgomaster.

9 Basson, *Philosophiae naturalis adversus Aristotelem libri XII. In quibus abstrusa veterum physiologia restauratur, & Aristotelis errores solidis rationibus refelluntur.*

10 Scaliger, *Exotericarum exercitationum liber XV. de subtilitate, ad Hieronymum Cardanum.* Note that this fifteenth book is the only one that Scaliger published, as there are no others!

11 For a good introduction to the controversy between Scaliger and Cardano, see Maclean, "Interpretation of Natural Signs."

12 See below, p. 101.

13 See the Testament of David van Goorle, MS. Hystoarysk Sintrum Ljouwert, Toegang 178, Martena e.o. MR 93.

14 See below, page xyz.

15 Jaeger, "David van Goorle," 220n1.

16 Universitätsbibliothek Basel, shelfmark Frey-Gryn. J VII 7.

17 See *Adresboek Nederlands drukkers*, 76.

18 Hooykaas, *Rheticus' Treatise on Holy Scripture*, esp. 18-19.

19 *Ibid.*, 14n8; Westman, *Copernican Question*, 130-31.

20 Westman, *Copernican Question*, 131.

21 See *ibid.*, 172. However, it made its way into a few historical surveys of the history of astronomy, such as Weidler's *Historia Astronomiae* of 1741 (where it is discussed, at p. 493, under the year 1651 as an "excellent letter").

22 On the phenomenon of the Descartes craze, see Des Chene, "Cartesiomania." See also below, p. 156.

23 Gorlaeus, *Idea physicae*, "Typographus Lectori Benevolo": "Uterque antehac editus nunquam, intra privatos delituisset parietes, nisi viri praeclari opera e tenebris erutus, dignus habitus esset, qui per manus studiosorum volitaret."

24 Gorlaeus, *Exercitationes*, XIII.i.343: "Has olim ad locos SS. Scripturae dabamus responsiones ..."

25 Gorlaeus, *Idea*, XIII.§9.47: "Galaxia non est Meteoron, sed minutissimarum stellarum splendor, quae ob nimiam parvitatem videri nequeunt: id quod se beneficio perspicilli nuper inventi observasse testatur Mathematicus quidam Patavinus." Cf. *Exercitationes*, XVI.ii.307: "Coeterum viam lacteam non esse partem coeli, sed minutissimarum stellarum splendorem, quae ob exiguitatem non ita radios spargunt, & ob propinquitatem splendorem quondam unum faciunt, beneficio perspicilli noviter inventi observatum est."

26 Gorlaeus, *Exercitationes*, I.i.1: "Quid sit philosophia." The manner of referencing to passages in the *Exercitationes* and the *Idea physicae* is explained in ch. 1, footnote 31, above.

27 Sennert, *Epitome naturalis scientiae*, "Disputatio prima, de natura philosophiae," tenet XI: "Philosophia est habitus intellectus, Sapientia et prudentia constans, omnia scilicet contemplans, atque actiones humanas congruenter gubernans, ut hinc homo summum bonum adipiscatur."

28 Gorlaeus, *Exercitationes*, I.i.4: "nuda entium cognitio."

29 *Ibid.*, I.i.5: "Philosophia est doctrina de perfectione animae humanae in hac vita."

30 *Ibid.*, I.ii.6-8.

31 *Ibid.*, I.iii.9-12. As will be explained in chapter 3, with this equation of ontology ("the science of being") with a type of "logic," Gorlaeus parts company with those among his friends who adhered to Ramism and who, following Peter Ramus, sought to develop a new dialectics to replace both metaphysics and logic.

32 See below, p. 84.
33 McKnight, "Francis Bacon's God."
34 Passmore, *Man's Responsibility*, 19. I owe this reference to Doina-Cristina Rusu.
35 Voetius, *Selectarum disputationum pars prima*, 4: "Ratio humana non est prior, notior, certior fide; ergo non est ejus principium."
36 Gorlaeus, *Exercitationes*, I.iii.14: "Verum Theosophia de Dei natura et eius attributis."
37 See below, pp. 120-22.
38 Gorlaeus, *Exercitationes*, I.iii.12, 14-5, 16.
39 Reif, "Textbook Tradition," 20. At Leiden, for example, one much relied on Cornelius Valerius' *Physicae seu De naturae philosophia institutio*, whose first chapter (*Quid natura*, 8) offered the following definition: "Natura definitur principium motus & quietis rerum corporearum: seu caussa, ut moveatur vel quiescat, eodemque in statu oliquamdiu permaneat id, cui inest primo, ac per se tanquam pars substantiae eius, non autem fortuito, vel, ut vulgus loquitur, per accidens."
40 Gorlaeus, *Idea*, I.§8.4.
41 Gorlaeus, *Exercitationes*, I.iii.13: "prima aut universalis philosophia."
42 *Ibid.*, I.iv.16-20.
43 *Ibid.*, II.i.25: "Sed varium est pro horum entium unione. Nonnunquam haec se invicem confuse contingunt; ut in acervo lapidum: aliquando ordine; ut in mundo: quandoquoque habent easdem qualitates, ut etiam ejusmodi unio sit inter illa; sicut inter hanc et illam aquae guttam: sunt quoque, ubi unum est in alio intime, illudque penetrat, et per illud agit; sicut anima in corpore. Quomodocumque plura entia inter se jungantur, totum illud semper erit ens per aggregationem. Neque enim unio mutat rerum essentiam, ut ex duabus rebus possit fieri una res numero."
44 Aristotle, *De generatione et corruptione*, 328a13ff.
45 On this episode, see Verbeek, "*Ens per accidens*"; idem, *Querelle*, "Introduction"; as well as below, pp. 148-50.
46 Pasnau, *Metaphysical Themes*, 597.
47 Gorlaeus, *Exercitationes*, XII.[i].222-23: "Ita in homine datur anima, datur quoque corpus, quae duo unita sunt, ita ut corpus sit animae domicilium, vehiculum et instrumentum, per quod anima suas exerceat operationes, sed ista duo non sunt unum ens facta, quod homo vocatur, sed unumquodque suam retinuit essentiam completam et perfectam, qua illud est, quod est. Homo vero non est idem quod anima, neque idem quod corpus, sed est idem, quod anima et corpus simul sumta et aggregata. Interim si homo non ut ens per aggregationem, sed per se unum considerandus sit, erit idem, quod anima existens in corpore." I have here followed Pasnau's elegant translation in *Metaphysical Themes*, 598.
48 Gorlaeus, *Exercitationes*, XII.[i].224: "... cupio dissolvi et esse cum Christo. Si enim compositum sit unum ens, et homo sit compositum, per mortem necesse est interire essentiam hominis, quia interit illud ens, quod est compositum. Quomodo igitur cum Christo esse potest, qui non est? Aut quomodo mors dici potest dissolutio, si sit substantialis corruptio? Non enim interire dicitur, quod tantum dissolvitur. Haec quum plane non cohaereant, valedicamus nugis hisce Peripateticis, et sanctam sequamur veritatem."
49 See above, pp. 12-13.
50 Pasnau, *Metaphysical Themes*, 678-79.
51 Gorlaeus, *Exercitationes*, XIV.[i].256: "Nego plane ullum corpus factum esse, nisi per solam creationem, quum Deus Optimus Maximus mundum hunc crearet: nego ullum interijsse, aut interire posse, nisi ab eodem in nihilum redigatur: nego unum corpus in aliud transmutatum esse, et transmutari posse." Pasnau, *Metaphysical Themes*, 705, adds: "Gorlaeus's atomism later has as its counterpart the monism of Conway and Spinoza... ."
52 Gorlaeus, *Idea physicae*, II.§3.6-§6.9.

53 Valla, *Repastinatio philosophie et dialectice*. On the substantial differences between Ockham's and Valla's ontology, see however Nauta, "Ockham and Valla."

54 Peeters, "Bij de heruitgave," 31.

55 Gorlaeus, *Exercitationes*, VII.vi.146: "... proferamus nostrum dogmatum Achilleum propugnatorem." Lasswitz, *Geschichte der Atomistik,* 1: 455-57, offers a good summary of Gorlaeus' nominalist arguments; see also Gregory, "David van Goorle," 46-49.

56 Gorlaeus, *Exercitationes*, IV.i.77, section title: "Nulla dari universalia." *Ibid.*, I.i.78: "Totum illud, quod in re existit, unum numero est, & singulare." *Ibid.*, IV.i.79: "Ille autem conceptus, qui diversitatem hujus rei ab alijs repraesentat, individuus dicitur, quia rei est proprius, eamque significant ut illis singularitatis notis, hoc, hic, nunc."

57 On the changing meanings of the label 'nominalism' and its early modern identification with 'atomism' (another tag that changed its meaning across time), see Pasnau, *Metaphysical Themes*, 83-92.

58 Bacon, *De principiis et originibus*, M3r: "Principium autem non est Ens; Ens mortale non est Principium; ut necessitas plane invincibilis hominum cogitationes (si sibi constare velint) compellat ad Atomum, quod est verum Ens, materiatum, formatum, dimensum, locatum, habens Antitypiam, Appetitum, Motum, Emanationem. Idem per omnium corporum Naturalium interitus manet inconcussum et aeternum. Nam cum tot et tam variae sint corporum majorum corruptiones, omnino necesse est, ut quod tamquam centrum manet immutabile, id aut potentiale quiddam sit, aut minimum." See Manzo, "Francis Bacon and Atomism," 224.

59 *Ibid.*, II.iv.52: "Quum enim ens sit commune & Deo & creaturis, nihil in suo conceptu includere potest, quod non competat & illi & his."

60 *Ibid.*, II.iv.54: "... unitas, veritas, bonitas, existentia, localitas, durabilitas."

61 See below, pp. 120-22.

62 Gorlaeus, *Exercitationes*, II.[i].97: "Et sicut unitas non est distincta ab eo, quod est unum; ita nec quantitas atomi a suo atomo."

63 See Bruno, *De triplici minimo*.

64 Gorlaeus, *Exercitationes*, II.i.31.

65 Hattab, *Descartes on Forms*, 159 and 168-72, 168-169: "Gorlaeus goes on to distinguish two kinds of modes, which correspond more or less to Descartes' distinction between an attribute, and a mode in the strict sense. [...] As examples of attributes, Descartes lists existence and duration. Likewise, Gorlaeus identifies the first kind of mode as a mode of being. [..] Hence Gorlaeus' sense of 'mode' is much closer to Descartes' than to Suárez's." On the consequences of Descartes' denial of 'real qualities' and their replacement by 'modi', see Menn, "The Greatest Stumbling Block."

66 *Ibid.*, 185.

67 *Ibid.*, V.i.88: "Accidens autem est vilius & imperfectius ens, quam substantia, nec aliquid substantiale."

68 Gorlaeus, *Idea physicae*, III.§3.15: "[...] interim unum accidens potest aliud producere. A calore enim effici videmus calorem, raritatem, levitatem, etc."

69 On real accidents and Gorlaeus' specific definition of them, see Pasnau, *Metaphysical Themes*, 259-60.

70 Gorlaeus, *Exercitationes*, VII.i.100: "Sicut diximus quantitatem non differre a corpore quanto, ita porro dicimus nullas potentias, immo nullas proprietates distingui ab essentia rei. Et quamvis in definitione haud ponantur, non tamen inde infertur, quod in re distinguantur: sed hoc tantum, quod sola ratione nostra, ac modo concipiendi differant."

71 Pasnau, *Metaphysical Themes*, 520.

72 Gorlaeus, *Exercitationes*, XIII.i.244: "Forte figura atomi est quadrata. Hic enim nullum dabitur vacuum. Forte diversae atomorum figurae. Ut sit, figura atomi tam exigua est, ut sensu non capiatur, vix intellectu. Quare illa quaestionem in medio relinquamus." On early-modern definitions of atomic shapes, see Lüthy, "Invention of Atomist Iconography."

73 Gorlaeus, *Exercitationes*, VII.vi.144: "Non enim haec oriuntur ex atomorum congregatione, sed in ipsis atomis existunt. Nisi enim singulares atomi forent siccae, non totum siccum foret ..."

74 Gorlaeus, *Idea physicae*, XII.§6.68: "Raritas, densitas, levitas, asperitas similes modi sunt, non reales qualitates. ... Nihil igitur aliud sunt quam partium situs."

75 Gorlaeus, *Exercitationes*, VII.vi.144-45: "Siccitas competit cuilibet atomo in sese. At ariditas est, quæ oritur ob partium cohæsionem, & totius duritiem. Ideoque hæc est affectio totius aggregati, non entis per se. Atque hoc modo ex omnibus aridis adhuc arte chymica potest humidum aliquod extrahi, ex sicco non potest." Kubbinga, *Molecular Worldview*, xix, derides the idea that Gorlaeus' distinction constitutes a first step towards molecular notions, but unfortunately without adding arguments to his invectives. For Kubbinga, it is only with Beeckman's idea that atoms compose physical *homogenea* (secondary particles) as the lowest species (*infimae species*) that we are entitled to speaking of 'molecules'. While no one has ever claimed that Gorlaeus had a fully developed molecular theory in this sense, it remains true that his distinction between physical and chemical properties that belong to the atomic level and others that belong to atomic clusters implies the idea of a stratification of substantial levels, whereby each possesses its specific properties. Such a two-tier model, while forcefully present in Gassendi, is for example absent in Descartes, who tries to reduce all qualities to the ultimate, corpuscular level.

76 Gorlaeus, *Exercitationes*, XII.[i].211: "Satis ardua, satis difficilia, fuere haec de ente in genere et accidentibus, in quibus hucusque occupati fuimus."

77 *Ibid.*, XIV.[i].250: "Nullam putamus in rerum natura dari aut materiam aut formam, quas Peripatetici partes corporum statuunt."

78 *Ibid.*, XIII.i.226: "Quis enim non putaret tenuissimas illas vaporis partes esse indivisibiles?"

79 *Ibid.*, XIII.i.228:"... quae ratio cogit entia multiplicari absque necessitatem?"

80 *Ibid.*, XXX.i.241, 243-46.

81 Libavius, *Alchymia triumphans*, 159. Note that a copy of this book was in the possession of Adriaan Metius at Franeker. It is today kept at Leeuwarden's Tresoar.

82 See Descartes, *Principia philosophiae*, bk. 2.

83 Gorlaeus, *Exercitationes*, XVII.i.313-14: "Elementa dicuntur, ex quibus corpora mixta constant, & in quae tandem resolvantur. ... Mixta enim, quae intereunt in terram & aquam resolvi experimur. ... Nullum enim est mixtum, quod in ignem aut aërem resolvitur."

84 *Ibid.*, XVII.ii.318-20.

85 *Ibid.*, XVII.iii.326, 328: "... videnti et tangenti."

86 *Ibid.*, XVI.i.303: "Sed nos negamus: aërem esse elementum."

87 *Ibid.*, XVII.iv.330: "Non enim aër suas potest deponere qualitates secundas, & assumere alias."

88 *Ibid.*, XII.[i].222-23: "Unamquamquam partem putamus suam & habere essentiam ante compositionem, & retinere post eam, neque fieri unum ens numero, aut ex ijs fieri unum ens; sed eas uniri & misceri ut fiat unum continuum, quod sit unum ens per aggregationem, non per essentiam." Cf. *Idea*, IV.§6.24-§7.25.

89 Gorlaeus, *Exercitationes*, XIII.ii.248: "Corpora minima sunt atomi, quae varie commiscentur. Hae se mutuo debent contingere. Si namque se invicem haud contingant: quomodo inde fiet corpus unum?"

90 Gorlaeus, *Idea physicae*, VII.§11.42: "Atomorum requiritur mutuus contactus et unio, non quidem substantialis sed accidentalis. Unio haec est continuitas quaedam: quae oritur quando accipiunt qualitates ejusdem generis."

91 Aristotle, *De generatione et corruptione*, I, 10, 328b.

92 Düring, *Aristotle's Chemical Treatise*, 11.

93 Scaliger, *Exotericae exercitationes*, ex. 101: "Mixtio est motus corporum minimorum ad mutuum contactum ut fiat unio." The twelfth-century physician Bartholomew of Salerno had however proposed a definition that might be viewed as an ancestor of Scaliger's: "commixtio miscibilium

per minima sibi junctorum unio." Quoted from Jacquart, "Minima in Twelfth-Century Medical Texts," 49.

94 Gorlaeus, *Exercitationes*, XIII.ii.248, repeats this definition verbatim, however without citing Scaliger as his source. In his *Idea*, VII.§11.42, he somewhat adjusts the definition: "Mixtio est motus minimorum corporum, per contactum mutuum ut fiat unio. Per minima corpora intelligimus indivisibiles atomos." On the *fortuna* of Scaliger's definition, see Zubov, "Zur Geschichte," 178, and Lüthy, "An Aristotelian Watchdog."

95 The best exposition of the doctrine of *minima naturalia* is Murdoch, "Medieval and Renaissance Tradition." For the modern period, see also Zubov, "Zur Geschichte."

96 Gorlaeus, *Exercitationes*, XIII.i.237: "Ita enim minima guttula aquae dicitur aqua, & minima particula arenae dicitur arena."

97 For an excellent Renaissance discussion of the various explanations of the origin of the "form of the mixt," see Fernel, *De abditis rerum causis*, book 1. For the transformation of "substantial forms" into "corpuscular structures," see Emerton, *Scientific Reinterpretation of Form.*

98 This doctrine of the creation of the 'temperament' owed incidentally as much to Galen's explanation of the (humoral) temperament as to Aristotle himself.

99 Gorlaeus, *Exercitationes*, XIV.[i].258: "Sane non apparent actiones elementorum, nisi qualitatum, quae tamdiu in se invicem agunt, donec una exsurgat qualitas in mediocritate, quae non eget substantia conservatrice, sed ipsa talis manet, donec ab extrinsecis accidentibus, qualitatibus, aut intendatur, aut remittatur."

100 Gorlaeus, *Idea*, VI.§42-43: "Calor et frigus habent causae efficientis rationem in mixtione. Humidum et siccum, rationem Materiae. Atque ita oritur temperamentum. Hoc nihil est aliud, quam debita harum quatuor qualitatum proportio."

101 See Land, *Geschiedenis van de wijsbegeerte*, 117; or Gregory, "David van Goorle," 51, who describes Gorlaeus and Sennert as representatives of a historical "moment in which Aristotelian physics, though already criticized and abandoned, has not yet been replaced by a different 'sytem' of the world," namely by "an atomist-mechanicist conception."

102 Bruno, *De triplici minimo*, 10: "nobis vero vacuum simpliciter cum atomis non sufficit... ."

103 On Bacon's relation to Democritus' atomism, see Rees, "Atomism and 'Subtlety'"; Manzo, *Entre el atomismo y la alquimia*, esp. chs. 5 and 6.

104 Sennert, *Hypomnemata*; see Lasswitz, *Geschichte*, 1: 436-54; Michael, "Daniel Sennert"; idem, "Sennert's Sea Change"; Stolberg, "Particles of the Soul"; Newman, "Robert Boyle's Debt"; idem, "Experimental Corpuscular Theory"; Clericuzio, *Elements, Principles and Corpuscles*, 24-33; Lüthy, "Sennert's Slow Conversion."

105 Gorlaeus, *Exercitationes*, XIV.[i].257.

106 See Vermij, *Calvinist Copernicans*, ch. 6.

107 Gorlaeus, *Exercitationes*, XVI.i.297-98: "Opticorum dogma est; diversa corpora Diaphana specierum visibilium facere refractionem. Stellarum autem visibiles species, & per cœlum & per aëra ad nos usque deferuntur; quia in cœlo sitæ sunt, & tamen a nobis videntur. [...] Ac contrarium [to the assumption that the light of stars and other celestial bodies are perceived after having undergone refraction] docent Mathematicorum observationes, quibus demonstratur nullam dari refractionem, nisi aliquando in hac inferiore aëris regione ob vapores & halitus."

108 *Ibid.*, XVI.ii.310: "Praeterea cometas apparuisse in ipso caelo certissimae docent Mathematicorum observationes."

109 Gorlaeus, *Idea physicae*, VI.3.33: "Cælum ergo statuimus esse corpus continuum, nec per se moveri, sed quiescere. Ipsæ vero stellæ in ipso moventur libere sicut pisces natant in aqua aut potius sicuti nubes circumferentur in aëre."

110 On the late sixteenth-century revival of the fluid-sky model, see Palmerino, "Bodies in Water."

111 On Metius' stay with Tycho Brahe, see Christianson, *On Tycho's Island*, 322.

112 See below, p. 90.

113 Gorlaeus, *Idea*, XIII.§9.47: "Galaxia non est Meteoron, sed minutissimarum stellarum splendor, quae ob nimiam parvitatem videri nequeunt: id quod se beneficio perspicilli nuper inventi observasse testatur Mathematicus quidam Patavinus." Cf. *Exercitationes*, XVI.ii.307: "Coeterum viam lacteam non esse partem coeli, sed minutissimarum stellarum splendorem, quae ob exiguitatem non ita radios spargunt, & ob propinquitatem splendorem quondam unum faciunt, beneficio perspicilli noviter inventi observatum est."

114 Palmerino, "La fortuna della scienza galileiana," 62.

115 Gorlaeus, *Exercitationes*, XVII.vii.334: "At locos vero S. Scripturæ respondent, eam de hoc motu loqui, non prout in se est, sed prout a nobis concipitur."

116 See above; I would like to thank Gerben Wierda for pointing this possibility out to me.

117 See Vermij, *Calvinist Copernicans*; the quote is from 372. For the brief period 1600-1603, Johannes Murdison is singled out as the only natural philosopher at any Dutch university to have addressed the Copernican model, while rejecting it. Metius at Franeker must obviously be counted among those who knew both Tycho's and Copernicus' cosmological models very well.

118 Gorlaeus, *Exercitationes*, XVII.vii. 334-35.

119 See Verbeek, "*Ens per accidens*, and idem, *Querelle*, "Introduction."

120 Gorlaeus, *Exercitationes*, XII.[i].230: "... asserimus animam & corpus esse duo entia, quae unum sunt per aggregationem." *Ibid.*, XII.[i].224: "Haec quum plane non cohaereant, valedicamus nugis hisce Peripateticis, & sanctam sequamur veritatem." *Ibid.*, XIV.[i].274: "Nego enim omnino animam formam corporis dici posse."

121 See Verbeek, "*Ens per accidens*." For the passage in which Voetius makes the link between the Arminian crisis and Gorlaeus' doctrine of man as *ens per accidens*, see below, p. 150.

122 See P. Voetius, *Prima philosophia reformata*.

123 Gorlaeus, *Idea*, III.§4.16-17: "Omnis substantia quae facta est, immediate a Deo producta est, et quaecunque fit a Deo producitur, nulla a substantia creata. ... Quaecunque ergo substantia fit, a Deo fit, quae perit, a Deo in nihilum redigitur: quaecunque etiam fit ex nihilo fit." Cf. Gorlaeus, *Exercitationes*, XV.i.277: "Ex hisce de materia & forma ita dictis, necessario sequitur, omnia ex nihilo facta esse, & in nihilum interitura, quae interibunt."

124 Gorlaeus, *Exercitationes*, XV.i.280.

125 At least in some parts of his *Philosophia naturalis* (1621), Sébastien Basson also favors an atomist doctrine of time, which he however pursues precisely for the opposite objective: he wants to get rid of as many substantial forms as possible, including the vegetative and sentient forms, and with all their functions as secondary causes. In fact, he sometimes suggests that God is not only the primary cause, but the only cause of everything that happens in nature.

126 Gorlaeus, *Exercitationes*, XVIII.i.337: "Atque inter eas [quaestiones] principatum obtinet celebris illa controversia: a quo producantur animae, A Deone vel a parentibus?" *Ibid.*, XIII.i.338: "Huic argumento respondebamus nos ..." *Ibid.*, XVIII.i.340: "Huic argumento respondebamus..." *Ibid.*, XIII.i.343: "Has olim ad locos SS. Scripturae dabamus responsiones ..."

127 For a discussion of the confessional aspects of this issue, see e.g. Freedman, "The Soul (*anima*) according to Timpler," 806-7. I thank Davide Cellamare for drawing my attention to this article.

128 *Ibid.*, XVIII.i.337: "... sed non est novum porcos luto delectari."

129 *Ibid.*, XVIII.i.339: "Huis et nos respondere solebamus: animam a parentibus ex nihilo produci, accedente singulari Dei concursu, quo ille animam variis donis ornaret, et indifferentem parentum generandi potentiam ad hoc individuum determinaret."

130 *Ibid.*, XVIII.i.344: "... animas creari in ipso momento, quo fit conceptio, et corporibus infundi." *Ibid.*, 345: "Hic dandum est, si reliquia animalia producant animam, multo magis hominem. At nos negamus et illam animam generare, et asserimus omnem animam a Deo ex nihilo creari."

131 I thank Davide Cellamare for summarizing for me the pedigree of this notion.

132 *Ibid.*, XVIII.i.346: "Praecipue urgeri solet argumentum de peccato originali."

133 *Ibid.*: "Hoc [peccatum] enim in corpore solo esse non potest, quum sit caecitas in intellectu, et perversitas in voluntate, quorum corpus non est capax, sed sola anima. Quodsi anima a Deo creatur, et haec una creabunt ab eo, qui animam creat et Deus peccati auctor statuetur."

134 *Ibid.*, XVIII.i.347: "Animae vero actiones multae dependent ab hoc temperamento. Quare bonae esse nequeunt, quia per organum exercentur pravum."

135 *Ibid.*: "Quare a Deo est privatio omnium illorum donorum Adamo in primo statu collatorum: sed proclivitas ad malum, quae est prava quaedam dispositio, est in ipso corpore, & a parentibus trahitur, a quibus corpus."

136 *Ibid.*, XIII.i.352: "Interim quid de anima vegetante statuendum sit adhuc mihi non constat. De anima vero sentientis et rationalis ortu ita esse statuendum, nempe eas a Deo creari, jam demonstratum est. FINIS."

137 Gorlaeus, *Idea*, XIII: "De anima humana."

138 *Ibid.*, XIII.§§1-3.

139 *Ibid.*, XIII.§5.73: "Quamvis anima plura intelligat quam e sensibus habet, tamen ea ipsa non intelligit, nisi per similitudinem quandam cum iis quæ sentiuntur: et eatenus verum est illud: quicquid est in intellectu verum est fuisse in sensu: si enim sine limitatione proferatur falsum est. nam et illa principia nobis sunt innata: esse Deum: unum, non esse duo. Quare non prius fuere in sensu quam in intellecto."

140 *Ibid.*, XIII.§6.

141 *Ibid.*, XIII.§12.75: "Voluntas est quæ res intellectas vel persequitur vel aversatur, persequitur volendo et eligendo [...]."

142 *Ibid.*, XIII.§13.

143 *Ibid.*, XIII.§15.76: "Actio voluntatis est libera, et coacta voluntas non est voluntas. Libertas hæc consistit in indifferentia."

144 *Ibid.*: "Quae libertas meæ voluntatis non in eo consistit quod possum bonum velle et malum nolle, sed quod possum idem objectum velle vel non velle, non vero nolle. Est enim nolle enim velle, sed velle non. Sic Deus, quod bonum vult, id libere vult: interi non potest velle malum."

145 See on the vitalistic aspect of early modern atomism, Clericuzio, *Elements, Principles and Corpuscles*; Henry, "Atomism and Eschatology"; idem, "A Cambridge Platonist's Materialism"; idem, "Occult Qualities and the Experimental Philosophy"; Shackelford, "Seeds with a Mechanical Purpose"; Hirai, *Le concept de semence*.

146 Kangro, "Erklärungswert und Schwierigkeiten"; Meinel, "Early Seventeenth-Century Atomism."

147 Newman, *Atoms and Alchemy*, esp. 98-124 and 190-219.

148 On the difficulty of using Descartes' mechanical model in chemistry, see Joly, *Descartes et la chimie*.

149 For example, a number of these heterogeneous elements converged in the rehabilitation of the ancient atomist Democritus, as I have tried to document in my "Fourfold Democritus."

150 Aristotle, *De anima*, II.1 (412a21).

151 On this school of thought, see Sabra's recent and authoritative "Simple Ontology."

152 Toletus, *Commentaria cum quaestionibus*, 60 (bk. 2, ch. 3, q. 8): "Fuit enim sententia quorundam Araborum ... qui dicebant causas has inferiors nihil operari, sed Deum solum ad ipsarum praesentiam effectus facere: unde solus Deus ad praesentiam ignis comburit, ad praesentiam cuiusque alterius agentis operatur: ipsae particulares causae solum se habent, ut signa actionis divinae. Hoc recipit inter Theologos Gabr.4.ent.dist.i.q.i. contra quam sententiam disputat S. Thomas.2. contra gen. ... Fundamenta Gabri. sunt duo: alterum est. Frustra fiunt per plura, quae possunt fieri per pauciora: sed potest Deus se solo operari cuncti: ergo frustra ponitur cooperatio causarum particularium."

153 Basson, *Philosophia naturalis*, 227 [numbered 247 by error]: "Voluerunt, inquam, hi philosophi quod Plato, Deum solum causam principalem reliquis uti, ut instrumentis."

154 Nielsen, "A Seventeenth-Century Physician," 348.

CHAPTER 3

1 Lasswitz, *Geschichte der Atomistik* 1: 482.

2 *Ibid.*, 1: 333.

3 Jaeger, *David van Goorle.*

4 Jaeger, "David van Goorle," 209-16.

5 E.g., Van Goorle, *Prognosticatie*. His prognostications appear to be up-market products, and their annual appearance seems to testify to a certain commercial success. Some, but not all, of his prognostications are discussed in *Bibliotheca Belgica*, 3: 263-68. On the long-winded titles, the author introduces himself as "Maistre Pierre de Goore/Docteur en Philosophie et Medicine/et souverain Amateur de la science d'Astrologie" (and analogously in Dutch). Peter van Goorle seems to have remained Catholic; he is mentioned as late as 1600 as the godfather of a child baptized at Antwerp. Valentine (Valentijn) van Goorle, who published prognostications in the 1580s, may have been his son.

6 Peeters, "Bij de heruitgave," 32.

7 The dates given by Jaeger, "David van Goorle," 212, are not correct. He follows by Van Buchell, *Diarium*, 20: "Sept. 1570. Vendidit dominium post hoc tempus Ordinibus Hollandiae Valburgis Uenara 90 millibus florenorum, cum eius actionem haberet Johannes Hornius Boxtelliae baro, Gorlaeis intercedentibus." According to this dating, the brothers would already have left Antwerp in their teenage years. However, we know that they were still in Antwerp around 1680. I wish to thank Gerben Wierda for proving this point for me.

8 On the Count, see *Nieuw Nederlandsch Bibliografisch Woordenboek*, s.v.; *Allgemeine Deutsche Biographie*, s.v. "Neuenahr."

9 In a document of the deputies of the Province of the States-General of 17 May 1589, for example, David Gorlaeus' father is mentioned as the Count's treasurer: "... bij syn Gen. van Nyenaer, door syn thesaurier David van Goorle..."; quoted from Jaeger, "David van Goorle," 212n1.

10 Quoted from *ibid.*, 216.

11 On the domicile of the Gorlaeus family at Utrecht, see Peters, "Bij de heruitgave," 31.

12 Hamerster, *Nauwkeurige Verklaring*, 1: 55. The request was made on 26 May 1609; the certificate carries the date of 18 June 1610 ("eenen David van Gorle te Cornjum op 26 may 1609 en by den Hove geinterineert d. 18 juny 1610"). "Ondertusschen valt het verleenen van zulke Brieven van Legitimatie hier by ons zeldzaam voor, wetende ik niet meer dan een geval, dat het zelve geschiedt is, te weten aan eenen *David Gorlè te Cornjum, op d. 26. May 1609., en by den Hove geinterineert, d. 18. Juny 1610."* In his current research into Gorlaeus' family, Gerben Wierda is considering the possibility that David Gorlaeus may have been adopted.

13 *Kronyk van het historisch genootschap te Utrecht*, 5th series (1865), 505-16; 502-5, quoted in Jaeger, "David van Goorle," 213n2.

14 *Ibid.*: "... ten insiene dat dselue Van Goorle is een lidtmaet onser landtschappe, deur dyen hy getrout hebbende d'oudste dochter vande here Doco van Martena, een vande voernaemste edelluiden ende alsnu mede een Staet deser landtschappe ende onsen medebroeder in Raede..." Quoted in Jaeger, "David van Goorle," 214n2

15 See Strong & Van Dorsten, *Leicester's Triumph.*

16 See on this issue Fruin, "Tien Jaren," 26-27; and Jaeger, "David van Goorle," 215n2.

17 Peeters, "Bij de heruitgave," 32.

18 See the Testament of David van Goorle, MS. Hystoarysk Sintrum Ljouwert, Toegang 178, Martena e.o. MR 93.

19 The information on Doeke van Martena is taken from the *Nieuw Nederlandsch Bibliografisch Woordenboek*, s.v.; De Crane, "Het aloud geschlacht," 206.

20 "Erat proverbium de eo, scilicet quod comsumpserit bona trium nobilium." Quoted from De Crane, "Het aloud geslacht," 206.

21 Peeters, "Bij de heruitgave," 32.

22 On the tombs underneath the floor of the church of Cornjum, see De Walle, *Friezen uit vroegere eeuwen*, 144-48.

23 On Abraham van Goorle, see *Nieuw Nederlandsch Bibliografisch Woordenboek*, s.v., and Langereis, "De verzameling munten." I would like to thank Jaap van der Veen for drawing my attention to this article.

24 Van Goorle's voluminous collections were not above suspicion: a number of coins and gems (cut stones on rings) were believed by experts to be forgeries; and rumor had it that a goldsmith named Arnold Mursens produced these forgeries for him (see Langereis, "De verzameling munten," 95).

25 Jaeger, 220.

26 Jaeger, "David van Goorle," 216-18n4, cites from a series of church documents dating to the years 1603-1607, in which Doeke van Martena, his daughter Swob van Martena and David van Goorle figure as the local lords who pay the teacher's salary, contribute to the renovation of the church tower and, together with Cornjum's parson Arnoldus Tebbingh, take on a series of further responsibilities.

27 On the function of these panels, see Bergsma, *Tussen Gideonsbende*, 164-65.

28 Fungerius, *De puerorum disciplina*. On Fungerius, see Sybrandy, "Johannes Fungerius"; Wumkes, "De 'Sylva Carminum'."

29 See *Biographisch Woordenboek der Nederlanden*, s.v.

30 Van Engelen, *Album Amicorum*: "etiamsi omnes contra sentirent, veritas defendendum; ac de praeceptore judicandum uti de maximo inimico. Haec tibi perpetua lex esto." See below, p. 101.

31 See Galama, *Wijsgerig onderwijs*, 17-18, who compares Franeker's Statutes with those of Leiden, Utrecht, Groningen, Deventer and Harderwijk.

32 The subordination to theology was defined in the university's Statutes as follows: "Cum philosophicus coetus etiam pars esse debeat Ecclesiae Dei, omnes philosophiae professores puram doctrinam Evangelii, quam Ecclesia nostra profitetur, amplectuntor: et ita philosophiam docento, ne traducant publice vel privatim doctrinam Ecclesiarum nostrarum: nec serunto, aut probanto, aut defendunto profanas opiniones: sed tuentor pacem publicam Ecclesiae, amando eam et ministros ejusdem" (article 18; quoted from Dibon, *Philosophie néerlandaise*, 130). On the general history of teaching at Franeker, see Galama, *Wijsgerig onderwijs*; Dibon, *Philosophie néerlandaise*, 127-63. On Franeker's theological approach to knowledge, see Jensma, "Inleiding," 11-14. On its (limited) appeal to foreign students, see De Ridder-Symoens, "Buitenlandse studenten"; for its specific importance to New England Puritans, see Sprunger, "William Ames."

33 Adama, *Crellii Isagoges*, 16: "vestigia Praeceptoris." On Adama's interest in a Ramist reinterpretation of logic, see however Galama, *Wijsgerig onderwijs*, 39-47. Stellingwerff, *De constitutione logices*, "Preface," [s.p.]: "fumosarum opinionum Pontifex ille."

34 Fockema Andrae e.a., *Album studiosorum*, 1: 38. The entry reads: "David Gorlaeus, (Utrajectinus), phil."

35 Galama, *Wijsgerig onderwijs*, 229; Peters, *Davidis Gorlaei Idea*, "Voorwoord," 5.

36 Fockema Andreae e.a., *Album studiosorum*, 1: 38, mentions, under the year 1605: "Fredericus Stellingwerff, iur."

37 Auction Catalogue Auletius, Tresoar Leeuwarden, MS. 13-13, inv.nr. 191, fol. 28r-35v, cont. 22r-26v, edited by M.H.H. Engels, http://mpaginae.atspace.com/Auletiustxtimg.htm (consulted 11 November 2011). I owe this crucial reference to Arjen Dijkstra.

38 Faber, "Disputatio sexta": "Stridebat aeternus tuus ipse ille quem nosti intime David Gorlaeus Ultrajectinus."

39 Stellingwerff, *De constitutione logices*, 69: "En tibi, Lector, *Disputationem Scholasticam* ante biennium fere in Franequerensi Frisiorum Academia a me, dum privato exerceor collegio, elaboratam, ac jam ad instigationem juvenis *Davidis Gorlaei* intimi mei, qui his majora molitur, in lucem editam."

40 See, e.g., Stellingwerff, *Politycq discours*.

41 Having just registered at the university, Gorlaeus was new in town in 1606, when the auction took place. This is why the auction catalogue refers to him, in Latin and in Dutch, to "David Gorleus, met Veno wonen" or "David Gorleus, cum Venone." After the second day of the auction, he had bought a sufficient number of expensive books to enjoy name recognition, so that the auctioneer simply calls him "David Gorleus." See Auction Catalogue Auletius, Tresoar Leeuwarden, MS. 13-13, inv.nr. 191, fol. 28r-35v, cont. 22r-26v, edited by M.H.H. Engels, http://mpaginae.atspace.com/Auletiustxtimg.htm (consulted 11 November 2011).

42 Much of the material presented in this section has been published previously in Lüthy & Spruit, "De Veno."

43 Vriemoet, *Athenarum Frisiacarum libri II*, 113; Boeles, *Frieslands Hoogeschool*, 1: 75; Galama, *Wijsgerig onderwijs*, 77.

44 Fockema Andreae e.a., eds., *Album studiosorum*, 1: 16: "Henricus de Veno, phil et ling et theol."

45 Meursius, *Athenae Batavae*, 351: "quorum nomina in historia non habentur." However, on Trutius and the philosophical teaching of the first decades of Leiden University, see Dibon, *La philosophie néerlandaise*, esp. 12-57. The theses that De Veno was made to defend have been used to document the traditional Aristotelian spirit reigning in the philosophical faculty in the first years of Leiden University; see Galama, *Wijsgerig onderwijs*, 78; Sassen, "Grotius," 39: "L'enseignement philosophique à Leyde était d'un dogmatisme exclusif."

46 See the statistics of foreign universities visited by Frisian students in Zijlstra, *Het geleerde Friesland*, 19-59.

47 The *Album* of Franeker University states, for example: "Anno 1609 rectore magnifico Henrico de Veno iuris utriusque, medicinae et philosophiae doctore, ethices ac physices professore" (Fockema Andreae e.a., eds., *Album studiosorum*, 43). The epithet "thrice great" is found, for example, in a student disputation of 1604: "D. Henrico de Veno, Phil. M. et I.V.D. Trismegisto, theologo insigni, liberalium artium magistro, ac in eadem academia physices ethicesque professor pectatissimo" (De Veno, *Disputationum physicarum octava*, dedication).

48 For the documents relating to De Veno's arrest and trial, see Lüthy & Spruit, "Doctrine, Life, and Roman Trial," as well as Baldini & Spruit, eds., *Catholic Church and Modern Science*, I/3, 2447-58.

49 See Prosperi, "Per la storia dell'Inquisizione," esp. 53.

50 "Henrici Veni Leovardiensis Frisii carcerati in sancto officio lecto eius processu in quo fatetur tenuisse hereses Calvini usque ad 18. annum, abinde citra, cum suae sit etatis annorum 23 asserit destituisse hereses. Decretum quod aliqui probi religiosi suae nationis cum eo agant, ut veritatem integre fateatur, quoniam benigne secum agetur." ACDF, SO, *Decreta*, 1597-1599, 113-13; transcribed in Lüthy & Spruit, "Doctrine, Life, and Roman Trial," 1121n28; Baldini, & Leen, eds., *Catholic Church and Modern Science*, I/3: 2453.

51 ACDF, SO, *Decreta*, 1598, fols. 241v-242r; transcribed in Lüthy & Spruit, "Doctrine, Life, and Roman Trial," 1121n31; Baldini & Spruit, eds., *Catholic Church and Modern Science*, I/3: 2455.

52 ACDF, SO, *Decreta*, 1598, f. 291v; transcribed in Lüthy & Spruit, "Doctrine, Life, and Roman Trial," 1122n32; Baldini & Spruit, eds., *Catholic Church and Modern Science*, I/3: 2456.

53 ACDF, SO, *Decreta*, 1598, f. 332r; transcribed in Lüthy & Spruit, "Doctrine, Life, and Roman Trial," 1122n33 and n34; Baldini & Spruit, eds., *Catholic Church and Modern Science*, I/3: 2458.

54 The lists are also cited in Firpo, *Processo*, 224 (n. 50c) and 306 (n. 50b).

55 Harrison, *Beginnings of Arminianism*, 22.

56 De Veno also signed the general register of Basel University (during the rectorship of the famous botanist Caspar Bauhin) in 1598-99 as "Henricus de Veno, Frisius." See Wackernagel et al, *Die Matrikel*, 2: 469, no. 54.

57 For a list of foreign universities at which Frisian students went to study, see Zijlstra, *Het geleerde Friesland*, 33; Bots and Frijhoff, "De studentenpopulatie," 59.

58 Vriemoet, *Athenarum Frisiacarum libri II*, 115; Boeles, *Frieslands Hoogeschool*, 1: 75; and Galama, *Het Wijsgerig Onderwijs*, 77, all quote the deliberations of the Senate of 18 September 1601, which give the reasons why Junius was preferred to De Veno: "hoewel zij [*sc.* the members of the Academic Senate] op persone van De Veno niet vele hadden te seggen, anders dat hij een jonghman was, die hem principalijcken in jure ende Medicinae geoeffent, ende noit geen *specimen* in Theologia [...] g' exhibeert hadde, ende daeromme soo vruchtbaerlijcken deselve professie niet soude cunnen bedienen, als de vorss. Junius." Incidentally, Junius did not heed the call to Franeker.

59 Vriemoet, *Athenarum Frisiacarum libri II*, 115.

60 See, for example, the funeral oration in honor of Frisia's state historiographer and Franeker's professor of eloquence, Pierius Winsemius, which recalls a physics disputation skillfully defended by the deceased under the supervision of "that great Henricus de Veno." See Wybinga, *Laudatio funebris*, fol. b2v: " ... in Physica magnum illum Henricum de Veno, I.V. et Medicinae Doctorem, Liberalíumque Artium Magistrum..."

61 See the Register of the Academic Senate of 18 December 1609 and 15 January 1610, cited in Boeles, *Frieslands Hoogeschool*, 1: 76, and Van Nienes et al, *De archieven*, 194.

62 Vriemoet, *Athenarum Frisiacarum libri II*, 117.

63 Boeles, *Frieslands Hoogeschool*, 1: 76; Galama, *Het wijsgerig onderwijs*, 76; the quote is from Napius & Lindeboom, 41: "... twisten ... die aan de Academie te Franeker woedden tusschen de aanhangers en tegenstanders van de leer van Aristoteles."

64 Harrison, *The Beginning of Arminianism*, 176.

65 This issue will be explained in detail below, pp. 109-10.

66 Van der Woude, *Sibrandus Lubbertus*, 127.

67 *Ibid.*, 183.

68 Van Limborch, ed., *Praestantium virorum epistolae*, 8.

69 On the Arminian issue, see below, pp. 105-13. On Lubbert's role in this affair, see Van der Woude, *Sibrandus Lubbertus*, 203-26. Reports of Episcopius' sojourn at Franeker are contained in Van Limborch, ed., *Praestantium virorum epistolae*, letters 131 (to Arminius) and 136 (to Corvinus).

70 Vorstius, *Anti-Bellarminus*, "Epistola." On De Veno's views on this issue, see below, pp. 90-92.

71 Letter to Johannes Becius of 27 September 1606, in Van Limborch, *Praestantium virorum epistolae*, letter 94, p. 176: "Nos dissensum metuimus. Sed quid si ea tenderet ad majorem consensum?"

72 Lubbert, *Responsio*, 2: "Sed haec libertas non vagabitur in infinitum; alioquin in foedem licentiam transformabitur. Quod igitur? Semper se continebit intra analogiam fidei." See Lubbert, *Declaratio responsionis*.

73 Van Berkel, "Franeker," 426-27.

74 According to the Records of the Academic Senate, 28 January 1611, De Veno was asked to "lessen ende doceren *horâ pomeridiana moralem* ofte *naturalem philosophiam Aristotelis*, ende hem soe in *docendo* als *disputando* wachten van subtile parerges ende quaestiën, oock van contumeliose daden ende woorden"; and that he had to "holden ende helpen onderholden *tranquillitatem academicam*, ende hem waachten van eenige correspondentie t'holden met studenten, het sy in de burse ofte daer buijten." See Boeles, *Frieslands Hoogeschool*, 1: 76-77.

75 Most, but not all, of the disputations defended under De Veno are listed in Postma & Van Sluis, *Auditorium*, 43. The problems surrounding the authorship of early modern university disputations

are notorious. (See on the issue of the attribution of disputational positions Rother, *Philosophie.*) However, in De Veno's case, the criterion of coherence, both among the doctrines expounded in the various disputations defended under his chairmanship and among the authorities invoked in them, leads us to regard these disputations (with the exception of the one metaphysical disputation *pro gradu*) to be a direct reflection of his teaching. We therefore assume that he either directly wrote, or at least approved of, the theses defended in the disputations over which he presided.

76 De Veno, *Syllabus errorum et contradictionum.* The only extant copy, which Arjen Dijkstra has thankfully managed to locate, is held at the New York Public Library (shelfmark SEB p.v. 20, no. 3). The syllabus is signed "H. de Veno" at the end. A dedication in De Veno's hand is unfortunately partly cut off.

77 On the development of a specifically Protestant metaphysics, see above all Leinsle, *Das Ding und die Methode.*

78 De Veno, *Disputationum prima*, thesis 1. Cf. Gorlaeus, *Exercitationes*, I.ii.6-9: "De animae perfectione." Note that Gorlaeus does not explicitly link our mental limitations to original sin.

79 De Veno, *Disputationum prima*, thesis 18. On seventeenth-century textbook definitions of natural philosophy (or physics), see Reif, *Natural Philosophy*, and "Textbook Tradition," 20.

80 De Veno, *Disputationum prima*, Corollaria, no. 1: "Quoad materiam, Physica Arist[otelis] non est perfecta."

81 In the prolegomenon of his *Cosmopoeia*, Casmann rebuts the arguments formulated in the sixth century by the Neoplatonic commentator Simplicius against the biblical account of creation. De Veno, *Disputationum prima*, thesis 24, mentions these arguments and states: "Quae autem hic adversus Mosen a Symplicio fabricata sunt, ut impia prorsus et pagana execramur et detestamur. Legi autem potest eorum refutatio apud Otthonem Casman. In proleg. Cosmop." On Casmann's principle "Cedat Aristoteles Mosi," see Mahnke, "Rektor Casmann," 330. On Casmann's relation to other "Mosaic philosophers," see Blair, "Mosaic Physics."

82 De Veno, *Disputationum secunda*, thesis 8.

83 *Ibid.*, questions 2 and 5.

84 De Veno, *Disputationum quarta*, theses 2, 4, 10, and 13.

85 *Ibid.*, thesis 18: "Nos cum doctissimo et subtili Scaligero locum hoc modo definimus: locus et spatium rei locatae, vel locati corporis, quod intra superficiem corporis exterioris ambientem continetur."

86 The rejection of Aristotle's "place" (*locus*) and its substitution with "space" (*spatium*) is found in Scaliger, *Exercitationes*, 15, in exercitatio 5, section 2: "Vacuum quomodo detur": "At nos illud profitemur vacuum, in quo corpus est. Idemque esse vacuum, et locum: neque differre, nisi nomine. Sane si non esset vacuum, non esset locus. Est enim vacuum, spatium, in quo est corpus." And section 3: "Loci definitio": "Non est igitur locus, exterioris corporis ambiens superficies: sed id, quod intra eam superficiem continetur."

87 De Veno, *Disputationum quarta*, thesis 18, question 3. For the confessional reasons why late-sixteenth- and seventeenth-century Calvinists tended to accept Scaliger's replacement of *locus* by a general *spatium* or *ubi*, and why they viewed "quantity" as an inseparable aspect of body, see Leijenhorst & Lüthy, "Erosion of Aristotelianism."

88 De Veno, *Disputationum quinta*, thesis 7.

89 *Ibid.*, theses 15-23.

90 De Veno, *Disputationum octava*, thesis 3: "Elementa sunt essentiae corporeae, specie individuae, mutationi obnoxiae, et in quas omnia mixta et constituuntur, et resolvuntur."

91 *Ibid.*, thesis 11.

92 *Ibid.*, thesis 9.

93 *Ibid.*, thesis 15.

94 *Ibid.*, theses 11, 24, 25.

95 *Ibid.*, theses 15, 18.

96 *Ibid.*, thesis 20.

97 De Veno, *Theoremata physica*, thesis 5.

98 *Ibid.*, thesis 17.

99 *Ibid.*, thesis 33: "Haec de aere et ejus regionibus, sufficiens sit contra multorum Aristotelicorum opiniones disputatio."

100 Aristotle, *Meteorology*, 328a5. Translation by Düring, *Aristotle's Chemical Treatise*, 41.

101 See Cardano, *De subtilitate*, 135: "Tria tantum vere in mistis terra et aqua pro materia, et calor colestis agens." See also Cardano, *In calumniatorem*, 1301 and 1296: "Tria sunt principia mistorum sed duo tantum praebent qualitatem, scilicet coelum et aqua," as earth has no quality: "Sola enim terra est expers omnis qualitatis, et tamen non est necessarium elemento ut habeat qualitatem, qui non concurrit ad ullam actionem ...". Piccolomni, *De mixtione*, fol. 108r, summarizes this doctrine as follows: "Affirmavit Cardanus in liber de mistione, mixtum non constare ex igne, nec ex aere, sed tantum ex terra, aqua, et celesti calore," that is, natural (mixed) bodies consist of two elements plus a quality, namely earth, water and celestial heat. On Cardano's natural philosophy and his theory of the elements, see Ingegno, *Saggio*, ch. 6, esp. 223-40. On Cardano's and Scaliger's debt to the Paduan commentary tradition on Aristotle's *Meteorology* IV, see Lüthy, "An Aristotelian Watchdog."

102 See above, pp. 44-46.

103 De Veno, *Quaestiones illustres*, thesis 10.

104 De Veno, *Disputationum prima*, thesis 40: "ut videre est in titulo libri 4. Meteorologorum qui vere non est meteorologicus"; corollarium 3: "Subjectum libri 4. Meteorologicorum est corpus perfecte mistum homogeneum."

105 On this issue, see Lüthy, "An Aristotelian Watchdog."

106 Lasswitz, *Geschichte der Atomistik*, 1: 335.

107 The complete title reads: De Veno, *Disputationum physicarum nona, de misti generatione et ejus interitu.*

108 *Ibid.*, thesis 3.

109 See Ingegno, *Saggio*, 234.

110 De Veno, *Disputationum nona*, thesis 7.

111 *Ibid.*, thesis 11.

112 *Ibid.*, thesis 12. Cf. Gorlaeus, *Exercitationes*, II.i.24-26, and *exercitatio* XIII.[i]("De atomis"), 225-49.

113 De Veno, *Disputationum nona*, thesis 14.

114 *Ibid.*, thesis 19.

115 See Fernel, *De abditis rerum causis*, bk. 2, ch. 10. This important treatise is now available in English translation, see Fernel, *On the Hidden Causes of Things*, ed. and transl. Forrester. This theory is also found in Cardano, *Contradicentium medicorum libri*, bk. 4, *controversia* 4, where the starting point is once more Aristotle's *Meteorology* IV, of which Cardano says: "Philosophus quarto Metheororum definit putredinem interitum calidi nativi in humido ab externo calore."

116 De Veno, *Disputationum nona*, theses 24, 25, 36, 37.

117 De Veno, *Theoremata de anima*, s.p., [i], prior to theorem 1: "erroneas veterum philosophorum de illa [*sc.* anima] opiniones."

118 *Ibid.*, theoremata 1-11, theorema 12: "Anima rationalis est forma informans hominem." Theorema 13: "Anima autem est forma substantialis hominis." Theorema 15: "Finalis animae causa sunt omnes operationes ipsius."

119 *Ibid.*, theoremata 19-23. On this question, De Veno refers to Thomas Aquinas and Javelli, relying for his unicist arguments on Mercenario. For a description of the pluralist and unicist views of souls and forms and their relation to early modern matter theory, see Michael, "Daniel Sennert," 275-86.

120 De Veno, *Theoremata de anima*, theorem 26.

121 De Veno, *Quaestiones illustres*. Incidentally, this is not only the earliest known master's degree defended at Franeker, but also the only extant master's disputation held between 1585 and 1613. See Fockema Andreae e.a., eds., *Album studiosorum*, 18.

122 De Veno, *Questiones illustres*, questions 2-4. De Veno refers here to Plato, Plotinus, Iamblichus, Seneca, Cicero, and to Foxius Morzillo's compendium of ethics of 1561.

123 *Ibid.*, question 6: "An subiectum metaphysices sit omne intelligibile quatenus tale, an vero ens qua ens? Prius Noeterici quidam, contra quos posterius sustinebimus."

124 Adama, *Theses logicae*, defended in 1606 the following position: "Res in dialectica considerata est ens et non ens, quod uno vocabulo cum D. Goclenio et Timplero, philosophis clarissimis, τὸν νοητόν, id est, omne intelligibile, rectissime significamus. Quicquid enim intellectu humano percipi et comprehendi potest, sive illud habeat essentiam, sive non, id usui logicae rectissime substernitur." On the influence of Ramism at Franeker, see Van Berkel, "Franeker als centrum." See also Clemens Timpler's Steinfurt metaphysics textbook of 1604, *Metaphysicae systema*, bk. 1, ch. 1, thesis 1, which opens with the definition of the subject matter of metaphysics as "omne intellegibile." On Timpler's metaphysics, see Freedman, *European Academic Philosophy*, ch. 11.

125 On the development of early modern Protestant ontology, see Leinsle, *Das Ding und die Methode*.

126 De Veno, *Quaestiones illustres*, questions 7 and 8. Like De Veno, many German and Dutch theologians and philosophers quoted from Bellarmine's *Disputationes de controversiis Christianae fidei* so as to illustrate the Catholic standpoint.

127 Incidentally, the use of Catholic and notably Jesuit authors at the Dutch Calvinist universities remained an uncomfortable issue. In 1610, the Leiden professor Franciscus Gomarus reproached his recently deceased colleague Arminius for having introduced his students to "the controversial and papist books of Thomas Aquinas, Suárez, Bellarminus and such vacuous polemicists" (*Bedencken*, 48). To this, Arminius' friend Petrus Bertius retorted "that all these books were already in use before Arminius' arrival, and concerning Thomas and Bellarminus: the late Junius [Arminius' predecessor, who died in 1602] himself recommended them to young students" (*Aenspraeck*, 3-4; the two quotes are taken from Cossee, "Arminius and Rome," 81). The uncertainty as to how to deal with Jesuit authors persisted through the century, and was characteristic of the so-called Calvinist scholasticism. Given his anti-Catholic rhetoric, the reliance of the orthodox Voetius on Thomas Aquinas, Suárez or on the Aristotle commentaries by the Coimbra Jesuits in his rebuttals of his Arminian opponents or of Descartes continues to be surprising.

128 De Veno, *Quaestiones illustres*, questions 9 and 10.

129 *Ibid.*, question 12. On Metius' work with Brahe, see Jensma, "Uit het huis," 459.

130 *Ibid.*, questions 13-17.

131 On Gorlaeus' rejection of the ether and the difference between the elements filling the sublunary and supralunary regions, see above, pp. 49-51.

132 De Veno, *Dissertatio politica*. This disputation is analyzed in some detail by Galama, *Het wijsgerig onderwijs*, 80-81.

133 *Ibid.*, corollaria, question 2: "An religio subditorum seu cultus Dei ad curam magistratus pertineat et an magistratus sit custos utriusque tabulae Decalogi? Affirmatur."

134 On Arminius' and Vorstius' positions on the relation between church and civil authorities, see above, 81-82, and below, 110, 116.

135 Timpler, *Disputatio metaphysica*, thesis 3: "Unde graviter errant Lutherani omnes, signum cum signato simul loco semper esse statuentes." I list this disputation under Timpler's name, not De Veno's, for reasons that will be explained below.

136 *Ibid.*, thesis 9: "Ergo necessario sequitur omne signum praeter rei conceptus esse ens externum et nullum internum. Hinc patet crassus ubiquitariorum error, qui in Eucharistia duplex signum statu-

unt, unum externum quippe pane et vinum, alterum internum, ut corpus et sanguinem Christi. Praeterquam enim quod corpus et sanguis Christi in Eucharistia sunt res signatae, male etiam signa appellantur, cum nullum signum praeter rei conceptum sit internum sed omne externum." Thesis 37: "Hinc nullo modo Signatum in signo esse potest .... Unde porro mnifestum evadit, graviter hallucinari eos qui in Eucharistia statuunt, corpus Christi esse in pane vel localiter, vel alio modo. Si enim signatum nullo modo potest esse in pane, cum illud sit signatum. Hic vero signum."

137 Timpler, *Disputatio metaphysica*. The Corollaries are followed up by this declaration: "Atque haec de generali Signi & Signati doctrina, ex Reverendo & Clarissimo viro M. Clemente Timplero, Praeceptore meo observando, hausta, breviter dicta sunt."

138 *Ibid.*, Dedication, A1v. Arnoldi's dedication goes, first of all, to the Count of Bentheim, the founder and patron of the Gymnasium Illustre. The list of the professors of the Steinfurt Academy, to whom this disputation is also dedicated, begins with "D. Conradus Vorstius SS. Theologiae Doctor."

139 On Casmann, see Mahnke, *Rektor Casmann*; on Timpler, see Freedman, *European Academic Philosophy*.

140 On the history of the Gymnasium Illustre, see Heuermann, *Geschichte des Gymnasium*, and Rübel, *Das Burgsteinfurter Gymnasium*. On its importance for the Netherlands, see Abels, "Das Arnoldinum."

141 See above, pp. 70-71.

142 The motto "Authoritates non habeo" is from a letter by De Veno to J. Saeckma of ca. 1597, repr. in Engels, ed., *Brieven,* 258, line 176.

143 See above, pp. 91-92.

144 See Leijenhorst & Lüthy, "The Confessionalization of Physics."

145 De Veno, *Disputationum quarta*, th. 18, qq. 2 and 3.

146 De Veno, *Quaestiones illustres*, th. 9: "An detur materia prima, eaque an sit substantia, an corpus, and incorruptibilis? Affir[matur]."

147 De Veno, *Disputationum prima*, th. 19: "Prima, nos ex ejus cognitione in Dei, ejusque potentiae cognitionem pervenire."

148 *Ibid.*, th. 8: "Execramur enim Academicorum opinionem tanquam indignam non solum Christiano, sed etiam Philosopho ..."

149 *Ibid.*, th. 23-25.

150 See above, pp. 52-53.

151 This assertion can be made by comparing De Veno's matter theory with that found in the disputations by Bertius, *De elementis*; id., *De mixtione*; Jacchaeus, *Disputationum physicarum octava*; A.A. Vorstius, *Theses physicae de elementis*; Murdison, *Theses philosophicae*, etc.

152 See Piccolomini, *De mixtione*, fol. 108r: "Affirmavit Cardanus in liber de mistione, mixtum non constare ex igne, nec ex aere, sed tantum ex terra, aqua et celesti calore... ." Cardano, *De subtilitate*, "Liber secundus, de elementis," 135: "Tria tantum vere in mistis terra et aqua pro materia, et calor coelestis agens"; also "Liber quintus de mistione et mistis imperfectis, sic metallicis." However, he remained not entirely consistent in his theory of the elements, see idem, *In calumniatorem*, 1031 and 1296: "Tria sunt principia mistorum, se duo tantum praebent qualitatem, scilicet coelum et aqua," the reason being that the earth has no quality: "Sola enim terra est expers omnis qualitatis, et tamen non est necessarium element out habeat qualitatem, qui non concurrit at ullam actionem... ." On Cardano's and Scaliger's debt to the commentary tradition on Aristotle's *Meteolory* IV, cf. Lüthy, "An Aristotelian Watchdog."

153 Cf. De Veno, *Disputationum nona*, th. 3: "Generatio est mutatio Elementorum a spiritu ad producendum corpus mixtum." *Ibid.*, th. 5: "Efficiens mixti principalis est Spiritus. Instrumentalis calor quilibet." Cf. idem, *Disputationum quinta*, th. 7: ... forma mundi ... gubernatur a nobilissimo Spiritu Dei."

154 De Veno, *Disputationum octava*, th. 11: "prima corpora generabilia & corruptibilia." Idem, *Disputationum nona*, th. 16: "Quare in errore eos versari manifestum est, qui aut formas elementorum in mixto manere integras tradunt ..." *Ibid.*, th. 22: "Quaeritur hic an mixti forma et ejus temperamentum differant? Aff. Quia substantia & acciens differunt."

155 Gorlaeus, *Idea*, VII.§1.43-44: "Corpora, quae ex terra et aqua miscentur non habent essentiam distinctam aut a terra aut ab aqua: sunt enim compositum quid. At nullum compositum esse aliud, quam suas partes, aliam habere essentiam, quam hae sunt, diximus ante."

156 Scaliger, *Exercitationes exotericae* (1576 ed.), ex. 11, sec. 1, p. 61; ex. 12. sec. 1, pp. 65-66; ex. 16, sec. 1, p. 79; ex. 20, sec. 1, p. 97; ex. 102, sec. 2, p. 373.

157 On the doctrine of *minima naturalia*, see Murdoch, "Medieval and Renaissance Tradition."

158 See above, p. 47.

159 Sennert, *De consensu ac dissensu* (2nd edn of 1629), 221: "Atque haec, quam proposuimus, est proculdubio antiquissimorum Philosophorum de mistione opinio, et ipsius Democriti, qui ex atomis res omnes componi et generationem nihil aliud, nisi σύγκρασιν et διάκρασιν, esse statuit [...]. Et procul dubio id sensit Scaliger, cum, loco allegato, scribit, mistionem esse motum corporum minimorum ad mutuum contactum, ut fiat unio."

160 Scaliger's corpuscular interpretation of minima was not unique. For example, in Buccaferrea's *Lectiones in quartum Meteorologorum librum*, 145, we find similar ideas. Indeed, Piccolomini, *De mixtione*, fol. 104r, attacked the view embraced by a whole group of Paduan colleagues to the effect that mixture was merely a "minimarum partium juxtapositio."

161 Scaliger, *Exotericae exercitationes*, ex. 16, sect 3: "Constantibus autem formis mistio acervus esset: quemadmodum etiam in siccis corporibus. .... Quae verae essent atomi Democriteae. Igitur essent certae quantitates, non tanquam misti partes, ut unius: sed suum sibi quodque totum."

162 Gorlaeus, *Exercitationes*, II.i.

163 *Album studiosorum*, ed. Du Rieu, col. 101, gives only part of the full entry in Leiden University's *Album studiosorum*, MS. Archives of the Senate and the Faculties, ASF7, where we read: "David Gorlaeus. Ultrajectinus. annorum XX. studiosus Theologiae – apud Magdalenam Laurentii filiam."

164 See below, p. 100.

165 Jaeger, "Van Goorle," 218.

166 Engelbert Egidius van Engelen, *Album Amicorum*, Leiden University Library, MS. Papenbroek, 21, f. 141, 90r.

167 Wierda, "Twee studiegenoten," 7.

168 See Wierda, "Twee studiegenoten," appendix 1: "Chronologie Engelbertus Aegidius Arnhemensis."

169 *Testimonium Academiae Ultrajectinae*, 28, extensive draft version of the *Corollaria* of the disputation defended on 18 December 1641 under Voetius (for the historical context, see below, pp. 148-50): "Assertio παραδοξολόγος Taurelli (quem Atheum Medicum vocabant Theologi Heidelbergenses, in judicio suo de Vorstii tractatu, de Deo, perscripto ad deputatos Synodi Hollandicae anno 1610), in *Triumpho Philosophiae*, in praemissis axiom. D. 4 & D. 5, quam imprudentia juvenili ex illo adoptare voluit popularis noster David Gorlaeus, tunc temporis in Theologia, cui studere coeperat, aut cujus studio destinatus erat, vertiginem patiens & vacillans, in Exerc[itationibus] Philosoph[icis] Exerc. 14. Pag. 267 qua statuitur, *Hominem ex anima et corpore compositum esse Ens & unum per accidens, non vero per se*; incurrit in veritatem non tantum Physicam (quam, Physicis explicandam reinquimus) sed et Metaphysicam, pneumatologicam, & theologicam. Monemus ergo studiosos nostros videant, ne uno absurdo temere dato, multa alia sequantur, & error parvus in principio, fiat magnus in fine."

170 Engelbert Egidius van Engelen, *Album Amicorum*, Leiden University Library, MS. Papenbroek, 21, f. 141. The discovery of this document is announced in Jaeger, "Van Goorle," 14-15.

171 For a number of reasons, this *Album* is a highly fascinating document. One aspect that has escaped the attention of historians is that it carries, in its intimate first pages (right after Engelbert Egidius' own title page and his own father's inscription, on fol. 3v-4r), the only entry by a woman. Everything about this inscription is conspicuous – first of all that its author writes in Spanish, French, Latin, Greek and Hebrew. The person who wrote this inscription was no one other than Juliana Morell (1594-1653), the celebrated Catalan girl who, resident in Lyon, obtained at age 12 her first university diploma "summa cum laude" in ethics and dialectics and at age 14 her doctorate in law. In fact, it was in the year of her doctorate, on 26 July 1608, that she signed off her name in Engelbert Egidius' *Album*, as "Christi Domini indignissima serva, Iuliana Morell Brachinonensis … die ac festo Beatae Annae 1608." Morell was Catholic and was later in the same year to enter a Dominican convent, taking the vows in 1610. That she should feature (and moreover in the personal section) of an *Album* that included otherwise only stern Calvinist divines is certainly surprising.

172 Quoted in Van der Kemp, *Kerkelijk leven*, 125: "… item dat ick die partiale namen van Arminianen, Remonstranten ende Contraremonstranten van de predigstoel sal laten."

173 *Ibid.*, 30.

174 On Engelbert Egidius van Engelen, see *Biografisch lexicon voor de geschiedenis van het Nederlands protestantisme*, s.v.; Van der Aa et al., *Biographisch Woordenboek der Nederlanden*, s.v.; Jaeger, "David van Goorle," 219; Brandt, *Reformatie*, 4: 2-4 and 9; Van der Kemp, *Kerkelijk leven te Arnhem*, 29-31, 126.

175 The printer, Johannes Jansonius, was also indirectly linked to the Arminian camp: his son, Johannes Janssonius, Jr., married into the Hondius family; his father-in-law, Jodocus Hondius, was also the father-in-law of Pieter van den Keere, a half-brother of Petrus Bertius, Arminius' friend and Leiden theologian and philosopher.

176 Martijn, *Christelijke bekeeringhe*, translator's dedication, s.p., dated 15 October 1615. On Jansson's publications, see Huiskamp et al., eds., *Catalogus van de pamfletten*; and Spaendonck, *Catalogus van de Arnhemse drukken*. As is the case with many other key publications treated in the present book, Jansson's Taurellus edition is exceedingly rare. More paradoxically even, given the role Taurellus' work played in the controversy surrounding Vorstius, neither the first nor even the second, Arnhem edition of Taurellus' *Philosophiae triumphus* are today to be found in any Dutch public library!

177 "No historian" – with the exception of Huib Zuidervaart, who told me about the fact that Van Echten lived in the same house as Gorlaeus, and Arjen Dijkstra, who first associated this name with the Socinian student from Franeker. I wish to thank both colleagues for providing me with this valuable clue.

178 Fockema Andrae e.a., eds., *Album studiosorum*, 46, student no. 1188: "Rudolphus ab Echten, nob(ilis), Drentinus, sch. Steinfurtensis, phil et art."

179 De Groot, "Franeker als Irenopolis," 113.

180 Sozzini, *De officio*. It has often been claimed that this publication was not much more than a prank (in the wake of Knuttel, *Verboden Boeken*, 84); but given that one of the students involved, Welsing, eleven years later published also a Dutch edition of the book (*Het ampt van een christen mensch*), it seems to me that the commitment to the contents of Sozzini's work clearly exceeded the level of a mere student provocation. As late as 1641, Gijsbert Voetius was still busy combating Welsing's work, together with others written by "Sociniaenen; Weder-doopers; Papisten; Remonstranten." See Voetius, *Catechisatie*, 620: "Welsing, lib. de Officio hominis Christiani Fol. D. 5. & 6."

181 Welsing had enrolled as a theology student on 28 June 1610; see Fockema Andrae e.a. eds., *Album studiosorum*, 47.

182 Omphalius had enrolled in philosophy on 29 October 1610; see *ibid.*

183 The quote is taken from a letter by Godefridus Sopingius to the theologian Johann Bogerman of 9 July 1611, quoted in Van der Woude, *Lubbertus*, 207. The description of the scandal and of its implications are from *ibid.*, 204-8, and from De Groot, "Franeker als Irenopolis."

184 We encounter Trebatius again as a new theology student, already aged 28, in the registers of Leiden University on 29 October 1612, see *Album studiosorum*, ed. Du Rieu, col. 108.

185 I shall here use as my terminology "Arminian" and "anti-Arminian," to avoid confusion. The former were also called "Remonstrants" and the latter "anti-Remonstrants" or "Gomarists" (after the anti-Arminian colleague of Arminius, Gomarus).

186 For the list of students who enrolled in 1611, see *Album studiosorum*, ed. Du Rieu, 39-40.

187 See e.g., Blockmans, "Formation of a Political Union," 129-40.

188 See e.g., Van Deursen, "Dutch Republic," 145-52,

189 The Greek verb προορίζειν, which would be translated as *praedestinare* ('to predestine'), is found in Acts 4:28; Rom. 8:29, 30; 1 Cor. 2:7; Eph. 1:5, 11.

190 *Catechis religionis Christinae*, 20-21: "54. *Quid credis de sancta et catholica Christi Ecclesia?* Credo Filium Dei, ab initio mundi ad finem usque, sibi ex universo genere humano, coetum ad vitam aeternam electum, per Spiritum suum et verbum, in vera fide consentientem colligere, tueri, ac servare: meque vivum eius coetus membrum esse perpetuo mansurum."

191 See Bakhuizen van de Brink, ed., *De Belijdenisgeschriften*, 53-54. Translation from Cossee, "Arminius and Rome," 84n36.

192 The summary of the life of Arminius is based on Bangs, *Arminius*; Muller, *God, Creation, and Providence*; Van Leeuwen, "Introduction"; and Stanglin, "Arminius and Arminianism."

193 See *Album Studiosorum Academiae Lugduno Batavae*, under 23 October 1576.

194 See, e.g., Cossee, "Arminius and Rome," 75.

195 Van Leeuwen, "Introduction," xi.

196 Cf. In Corvinus' *Christelicke ende ernstighe vermaninghe*, we encounter a succinct summary of the difference between the three main positions on predestination: according to Corvinus, Gomarus stands for the view that God's decree concerned uncreated man (supralapsarianism); the theologian Reinier Donteclock (against whom Corvinus argued in his book) stands for the view that the decree concerned man "created and fallen in Adam"; while Arminius stands for the view that the decree concernes man created, fallen in Adam, "as well as believing or unbelieving." Only according to the third view did God consider whether an individual decided for, and persisted in, faith.

197 That the two sets of disputations of 1604 were generally seen to have marked the beginning of the controversy can be seen from the fact that they were translated into Dutch and published jointly, cf. *Twee disputatien van de Goddeliicke Predestinatie.*

198 Arminius, *Opera theologica*, 283 (= Disputatio publica XV, ii): "Praedestinatio itaque, ad rem quod attinet ipsam est decretum beneplaciti Dei in Christo, quo apud se ab aeterno statuit fideles, quos fide donare decrevit, justificare, adoptare et vita aeterna donare ad laudem gloriosae gratiae suae." The translation is from Cossee, "Arminius and Rome," 76.

199 Van Leeuwen, "Introduction," xiv. Note that Muller, "The Christological Problem," emphasizes Arminius' refusal to view Christ as *autotheos* in the same ways as the Father was *autotheos*. In his lesser-known controversy with his colleague Lucas Trelcatius Jr., this Christological difference came clearly to the fore. According to Muller, this subordinationist tendency is linked Arminius' view of the role of Christ in the predestination issue.

200 The complete title of Johannes Wtenbogaert, *Tractaet Van t'ampt ende authoriteyt eener hoogher christelicker overheydt, in kerckelicke saecken* ("Treatise on the function and authority of a higher Christian government in ecclesiastical matters," 1610) renders the basic idea very well.

201 See *ibid.*, xiv-xv. Cf. also Cossee, "Arminius and Rome," 84.

202 Quoted from Bangs, *Arminius*, 299.

203 Arminius, *Verclaringhe.*

204 *Ibid.*, xv.

205 *Ibid.*, xvi.

206 *Ibid.*
207 Slatius [attr.], *De gepredestineerden dief.*
208 Van Leeuwen, "Introduction."
209 Cf. Bosch, *Petrus Bertius.*
210 Bertius, *Liickoratie*; Gomarus, *Bedencken.*
211 Muller, *God, Creation, and Providence*, 3. This triparition is based on Winder's *Comparative Darstellung*, part 2.
212 See, for example, Van Leeuwen et al., eds., *Arminius, Arminianism, and Europe.*
213 See Holtzapffel & Van Leeuwen, eds., *De Remonstrantie 400 jaar*, 14-19; and various encyclopedia entries, s.v. Remonstrance.
214 Leibniz, *Theodicy*, 58.
215 Dilthey, *Das natürliche System*, dedicates an entire chapter to "Der Rationalismus. Auflösung der Kirchenlehre durch Sozinianer und Arminianer." See also Daugirdas, "Biblical Hermeneutics"; van Slee, *Geschiedenis*, 132 and passim; Harnack, *Lehrbuch der Dogmengeschichte*, part III
216 See Daugirdas, "Biblical Hermeneutics," 92-93, and the literature quoted there.
217 Sozzini, *De auctoritate*, ed. Vorstius, "Praefatio apologetica," 5*r-v*: "... libellus, in gallicum sermonem jam conversus, a Theologis Basiliensibus, ut quidem Typographus istius loci testatur, post accuratam ipsius lectionem, plane fuisset approbatus: exceptis tantum tribus sententiis, quasi idem ipsi brevibus censuris, sive annotationibus [...] emendare studuerunt." Quoted after Daugirdas, "Biblical Hermeneutics," 93n17. On the relations between Calvinism, Arminianism and Socinianism, see also Rohls, "Calvinism."
218 Vorstius, *Anti-Bellarminus*, "Epistola dedicatoria."
219 Van Limborch, ed., *Praestantium virorum epistolae*, let. 175, p. 288: "Et illic quidem assensio stricte semper urgenda: hic vero libertas aliqua inquirendi, aut etiam dissentiendi, doctis omnino concedenda est; ne veritati, magisque insinuare se cupienti, ostium occludere velle videamur. Sancta, inquam, atque Christiana moderatio hic, ut in omnibus, servanda est: ne, dum unum praecipitium nimium vitare cupimus, in aliud non minus periculosum incidamus." For the pre-Spinozist history of the term *libertas philosophandi*, see Sutton, "The Phrase *Libertas philosophandi*"; for the link between Arminianism, tolerance and *libertas philosphandi*, see Simonutti, *Arminianesimo e tolleranza*, 15-42; Van Gelder, *Getemperde vrijheid*; Van Bunge, *From Stevin to Spinoza*, ch. 1.
220 Gualtherus, *De vita et obitu*, s.p. [65] and [80]. See also the telling title of Eglisemmius' pamphlet of 1612, *Crisis Vorstiani responsi. Qua D. Conrad Vorstius denuo Atheismi, Ethnicismi, Judaismi, Turcismi, Haereseos, Schismatis, et ignorantiae arguitur.*
221 Cf. James I, *Translaet Vanden Brief* (the Dutch broadsheet translation of the Letter by King James I of 6 October 1611).
222 Cf. Heuermann, *Geschichte des Gymnasium Illustre*, 91.
223 Fuller, *Church History*, X.4.
224 The pamphlets are listed in Knuttel, *Catalogus*, I, starting from p. 358. However, the dates given by Knuttel differ frequently from those given in Dibon, *Philosophie néerlandaise.*
225 Leiden's *Album studiosorum*, ed. Du Rieu, states on its list of professors, on p. ix, "Conrad Vorstius. Cathedram non adscendit." For different accounts, see the biographical literature cited in footnote 228.
226 *Judgement of the Synode Holden at Dort*, 102-3.
227 Harrison, *Beginnings of Arminianism*, 188.
228 On Vorstius' life, see Gualtherus, *De vita et obitu*; Baudartius, *Memoryen*, vol. 1, *passim*; Zedler, *Grosses Universal*-Lexikon, 50: 1290ff; Heuermann, *Geschichte des Gymnasium Illustre*, 75-94; Dibon, *Philosophie néerlandaise*, 80-84; Bangs, *Arminius*, 292, 322; Wenneker, "Vorstius"; Mühling, "Arminius und die Herborner Theologen," 130-33; Rohls, "Calvinism," 22-37.

229 In the *Requeste vande Studenten* of October 1610, Vorstius was accused of Socinian sympathies (Aiii*v*). The students argued that his appointment would lead to a probabilistic theology and a quodlibetal approach to truth to the detriment of the authority of the Reformed Church. According to this pamphlet, the submission had been signed by fifty-five students (Ai*r*), that is, "all theology students, except a few" (Aiv*v*: "Wy alle de Studenten der H. Theologie, (wonder weynig uytgenomen), inde Universiteyt van Leyden"). Incidentally, the date on the published version is 1610, but Knuttel, *Catalogus*, I, item 1872, corrects this date: "moet zijn 1611." The chronology of events given by Dibon, *Philosophie néerlandaise*, 82-83, makes 1610 however much more plausible.

230 See above, p. 116.

231 Vorstius, *Tractatus de Deo*, "Epistola dedicatoria."

232 Zedler, *Grosses Universal-Lexicon*, vol. 50, cols 1300-1307, lists Vorstius' most frequently attacked positions.

233 See Vorstius, *Tractatus de Deo*, notably disputation 3, plus explanations on 206-310.

234 Zedler, *Grosses Universal-Lexicon*, vol. 50, cols 1300-1307, tenets 4-18, with references to the passages in Vorstius' *Tractatus de Deo*.

235 The English list of heretical positions, together with Vorstius' rebuttals, are found in Vorstius, *Christiana et modesta responsio*; the quote is from p. 1: "Deus non est essentialiter immensus, nec simpliciter infinitus: sed est quantum, finitum, in loco, quodammodo corporeum, constans quasi ex materia et forma." The offending passages were located by Her Majesty in Vorstius' *Tractatus de Deo*, 202, 210 and 232-40.

236 Vorstius, *Tractatus de Deo*, "Epistola Dedicatoria," s.p. [i-ii]. The view that "recta ratio atque sensus sunt naturali fidei principia" is listed by Zedler, *Grosses Universal-Lexicon*, vol. 50, col. 1300, as the first among the twenty commonly cenured tenets. Tenet 2 states: "In definitione Dei ut genus licet analogicum, ita differentia dari potest a forma ejus petita" (cf. Vorstius, *Tractatus de Deo*, 153). Tenet 3 sustains that "Deus est substantia spiritualis; nullo nimirum sensu exteriore perceptibilis; licet tamen undem sensu generaliori, latiori & impropriori corpus dicere prout illud veram substantiam atque essentiam denotat" (cf. *ibid.*, 195).

237 Vorstius, *Tractatus de Deo*, "Ad lectorem," fol. 1*v*: "Multa hic quidem, fateor, agnoscenda sunt clara [et] certa ... primis ac communissimis illis sanae Philosophiae principiis, quae de Deo non minus, quam de aliis in genere vere entibus, aut substantiis, aut Spiritibus indubitate vera sunt, tanquam immotis fundamentis nituntur."

238 Gorlaeus, *Exercitationes*, I.iii.14: "Verum Theosophia de Dei natura & eius attributis."

239 See above, p. 99.

240 *Bedenckingen*, 8-9: "Nos editum dolemus non pauca quidem argute, et subtiliter de Deo, in eo dissecuntur, utinam vero, author cogitasset, illud Tertulliani, de Deo etiam verum dicere periculosum esse, hoc agree videtur, ne quid cum alijs commune sentiat, de Majestate Dei; non dubitat ergo, veterum et recentiorum Theologorum convellere doctrinam. Sola ea placent quae forte in lacunis Scoti, et athei illius medici Taurelli, invenit portenta. Deum essentia quantum, magnum, finitum, compositum ex essentia, et accidentibus, mutabilem voluntate, passivae obedientiae obnoxium esse et similia trecenta."

241 Taurellus, *Alpes caesae*, 2*r*: "Scheckianae philosophiae perpetuo fui studiosissimus."

242 On the life and work of Taurellus, see Baier, *Biographiae professorum medicinae*; Feuerlein, *Taurellus defensus;* Zedler, *Grosses Universal-Lexicon*, vol. 42, col. 401-2; Schmid, *Taurellus*; Mayer, *Taurellus*; Mayer, "Ein Altdorfer Philosophenporträt."

243 Taurellus, *Emblemata*, "Ad Lectorem," s.p., [i]: "Aristotelicae philosophiae et medicinae Galenicae professor." Idem, *Medicae praedicationis methodus,* "Praefatio": "Philosophus Christianus."

244 See Taurellus, *Synopsis Aristotelis metaphysices*. The double-truth argument is criticized in his *Philosophiae triumphus*, "Epistola dedicatoria, fol. 4*r*: "Primum enim consyderare coepi, duplex ne posset unius esse rei veritas, ut quod Theologice falsum est, verum possit esse Philosophis..." Cf.

also *Medicae praedicationis methodus*, "Praefatio," s.p.: "Idem enim Philosophiae Deus est & Theologiae."

245 Rudolph Goclenius, letter "Egregio philosopho Nicolao Taurello, amico suo," prefaced to Taurellus' *De rerum aeternitate*, s.p.: "Hoc, Taurelle, recte mecum judicas, sicut & istud: Erroneam esse sententiam eorum, qui existimant, Mundum esse aeternum, verum esse in Philosophia, in Theologia falsum: nec Aristotelem aliter judicare potuisse, quam mundum nunquam coepisse..."

246 Leinsle, *Das Ding und die Methode*, 1: 147.

247 Taurellus, *Philosophiae triumphus, hoc est, metaphysica philosophandi methodus, qua divinitus inditis menti notitiis humanae rationes eo deducuntur, ut firmissimis inde constructis demonstrationibus aperte rei veritas elucescat et quae diu Philosophorum sepulta fuit aucthoritate Philosophia victrix erumpat: quaestionibus enim vel sexcentis, ea quibus cum revelata nobis veritate Philosophia pugnare videbantur, adeo vere conciliantur, ut non fidei solum servire dicenda sit, sed eius esse fundamentum.*

248 Taurellus, *Philosophiae triumphus*, d6*r*: "Homo, corpus et anima. Homo non est unum per se, quod duabus immutatis constituatur formis. Accidens est hominis compositio, quod eius formae per se subsistere possint."

249 Ibid., 36: "Nos enim ex corpore et anima constituimur, sed voluntas per se consyderata simplex est animae facultas, quae sine corpore intelligi, atque consistere potest, qua ratione bonum quid existimanda est licet ob corporis affectus prae bono malum appraehendat."

250 Zedler, *Universal-Lexicon*, vol. 42, col. 401: "Insbesondere konnte er die Lehr-Sätze des Aristoteles von Gott, den Intelligentien, von der Vorsehung, und von der Seele, mit den Grund-Sätzen der Christlichen Religion keineswegs zusammen reimen."

251 On Taurellus' views on Ramus, see his letter of 19 October 1580 to Theodor Zwinger (Universitätsbibliothek Basel, MS Frey-Gryn. MS. II.4. no. 307). Cf. also Taurellus, *Alpes caesae* 37.

252 Even in those years when Taurellus was officially writing about other topics, eschewing metaphysics, he continued to air his anger at Aristotle's metaphysics at the most unlikely places (cf. e.g., Taurellus, ed., *Arnaldi Villanovi Opera*, col. 1691-92). See also Leinsle, *Das Ding und die Methode*, 1: 150n36.

253 Taurellus, *Philosophiae triumphus*, 87.

254 *Ibid.*, "Praefatio." Idem, *Kosmologia*, "Ad lectorem praefatio."

255 Vorstius, *Tractatus de Deo*, "Preface," 4a.

256 See the full title of Zanchi, *De natura Dei*.

257 Zedler, *Grosses Universal-Lexikon*, vol. 60, cols. 1520-33, provides a list of Zanchi's censored views. *Ibid.*, 50, cols. 1300-1307, provides Vorstius' list.

258 Taurellus, *Philosophiae triumphus*, th. 14 and 18 (p. a4*r*) defends man's free will and attacks the doctrine of predestination. Cf. also th. 27 (p. a6*r*): "Falluntur enim Theologi nihil homini tribuentes, ac si subiectum mere passivum essed, quidvis efficiente Spiritu."

259 Zeltner, *Historia crypto-Socinianismi*.

260 Leinsle, *Das Ding und die Methode*, 1: 156-57 concludes, for example, that Taurellus' works were on the whole not particularly influential.

261 Vorstius, *Christiana et modesta responsio*, 32-33.

262 See above, p. 102.

263 Taurellus, *Philosophiae triumphus*, 122: "unum et esse convertuntur"; "multitudo non est substantia." Cf. idem, *De usiis per se subsistentibus*, 2-3.

264 Leisle, *Das Ding und die Methode*, 1: 160. Cf. Taurellus, *Philosophiae triumphus*, d7*r*: "Singularia, et Universalia. Quicquid existit est singulare. In intellectu solum sunt universalia."

265 Taurellus, *Philosophiae triumphus*, 123, defines a *compositum* as "per se multa simplicia quae per accidnes composita sunt."

266 Taurellus, *Philosophiae triumphus*, 100; idem, *De rerum aeternitate*, 397, 408, 409, *et passim*.

267 Taurellus, *Synopsis metaphysices*, §55.

268 E.g., Taurellus, *Philosophiae triumphus*, d2*v*, th. 165.

269 Leinsle, *Das Ding und die Methode*, 1: 162.

270 Taurellus, *Synopsis metaphysices*, § 55: "In numero & magnitudine quaerendum est infinitum. In numero est τὸ ἕν. Quid in magnitudine? Atomus, punctum, τὸ κίνημα, καὶ νῦν. ... Omnem vero quantitatem esse dividuam, impudens est postulatum: de quo suo tempore, Deo volente, seorsim, prout res ipsa postulat, agemus. Nil compositum & finitum est, quod suis non constet principiis & primis, & simplicibus. Quo nomine Aristoteles saepe (atque utinam semper) progressum in infinitum ceu ἀδύνατον quidpiam explosit."

271 The reference to the two 'lost' works are given below, in footnotes 272, 276 and 277. Baier, *Biographiae*, 7, refers to Taurellus, *De infiniti continui sectione*, Frankfurt, 1597, but admits that he has found neither this work nor the *Commentaria in Hippocratis de natura hominis*. I owe the reference to the 1597 Leipzig book fair catalogue to Ian Maclean.

272 Taurellus, *Kosmologia*, 2: 145: "Atomos verum est magnitudinis esse principia. ... Nihil enim est majus: nihil magnum, quod non etiam primis, minimis, & individuis constet partibus: ut integro alias opere manifestum fecimus: quod nuncupavimus Apirotomen." The title *Apirotomen* refers to Taurellus' (unfindable) work *De infiniti continui sectione*.

273 *Ibid.*, 2: 119.

274 Taurellus, *Philosophiae triumphus*, 124, 170.

275 Taurellus, *Uranologia*, 2: 190: "Terra quidem, & aqua, corporum generandorum materia sunt. ... Ignis enim astrorum, maxime vero solis, vicarius, in intimis terrae partibus admirandorum naturae operum efficiens est causa primaria."

276 Taurellus, ed., *Arnoldi Villanovi Opera*, col. 8 ("commentarium"): "Sed vera haec est Medicorum sententia, nihil elementis prius et simplicius existere, quod nos libris ante citatis demonstravius." The marginal reference is to the *Philosophiae triumphus* of 1573 and to the lost or unpublished *Commentarii in libellum Hippocratis de Homine*.

277 *Ibid.*, col. 84: "De atomis alias egimus copiosius, easque demonstravimus non falsum esse commentum, sed dogma verum commentis obfuscatum falsissimus, e quibus & illud est, quod a Galeno secundum Hippocratis sententiam strenue convellitur, Atomos nimirum omnis esse qualitatis expertes." The marginal reference states: "In Commentarij in lib. Hipp. de humana natura. Lib. I. de Elemen. Cap. 2."

278 E.g. Taurellus, *Kosmologia*, 142-43: "Quod de atomis dicitur ex atomistarum sententia, nullius est momenti. Non enim atomi mundi, mundorumve condendorum materia sunt. Alia vero est materia, quam primam vocant peripatetici. Haec enim cum sit informis, & a nullo facta, etiam infinita est: ut ejus exigua duntaxat aliqua pars ad hujusce finiti mundi procreationem assumpta, & absumpta sit."

CHAPTER 4

1 Wierda, "Twee studiegenoten," 2. Gerben Wierda is currently conducting research on Gorlaeus' family connections in Rotterdam; in this context, he has developed the hypothesis that David might have died there. His findings will hopefully be published in due time.

2 See above, 54-56.

3 See above, 33-34.

4 Jaeger, "David van Goorle," 220, 229; Bosch, *Petrus Bertius*, 152.

5 Mersenne, *Quaestiones in Genesim*, col. 1838.

6 Mersenne, *La vérité des sciences*, 109.

7 Mersenne, *L'impiété des déistes*, 1: 237-38.

8 Quotes from Ariew, *Descartes and the Last Scholastics*, 126.

9 Naudé, *Advis*, 135: "En Philosophie, commencer par celle de Trismegiste qui est la plus antienne, poursuivre par celle de Platon, d'Aristote, de Raymond Lulle, Ramus, & achever par les Novateurs Telesius, Patrice, Campanella, Verulam, Gilbert, Iordan Bruno, Gassend, Basson, Gomesius, Charpentier, Gorlee, qui sont les principaux d'entre une milliace d'autres."

10 D'Espagnet, *La Philosophie*, Preface by Bachout, "Discours a la recommendation de la Philosophie ancienne restablie en sa pureté; Et sur le nom de son premier Autheur," s.p. [viii]: "L'Allemagne & l'Angleterre ont eu aussi plusieurs Autheurs qui n'onst suivy les opinions d'Aristote qu'aux endroits où il les ont treuvées les plus raisonnables, comme ont fait Bacon, Flud, Gorleus, Taurellus, Carpentarius & autres, dont quelques-uns ont escrit sur de nouveaux principes."

11 Cf. See on this theme Ribard, *Raconter, vivre, penser*, 334ff: "La *Science universelle* de Sorel, les livres de philosophie et la question de la rupture."

12 Sorel, *Science universelle*, 4: 438-40: "Il faut louer la grandeur du courage de Telesius, d'avoir osé le premier censurer les anciennes erreurs [...]. Patrice est aussi fort recommendable d'avoir détrompé son siècle, touchant beaucoup d'opinions absurdes des Corps célestes et des terrestres; Ces premiers, avec Cardan et quelques autres, [...] ont fait connaître que le vrai nombre des Élémens n'est que de deux. [...] à quoi s'accorde Gorlaeus et l'Auteur de l'Enchyridion. Copernic, Galilée, Iordan Brun, et Descartes, nous apprennent tout ce qu'on peut imaginer et supposer du nombre, de la sitation et du movement des Corps principaux de l'Univers; [...]."

13 Sorel, *Perfection de l'homme*, 209-10; 210: "Quoy que le nom de Novateur soit odieux à plusieurs personnes, il faut prendre garde que si en matiere de Theologie il est à aprehender, il ne l'est pas ainsi dans la Philosophie naturelle et humaine."

14 *Ibid.*, 238-43.

15 *Ibid.*, 248: "Suivant l'ordre des temps, nous viendrons à des Novateurs qui ont escrit en Latin, aussi bien que d'autres qui les ont precedez, & qui se sont servis des reigles de la Philosophie. Entres les Modernes de qui l'on peut faire quelque cas, il y a un David Gorlaeus Hollandois, qui a fait un livre apellé, *Exercitationes Philosophicae*, où il entreprend de combatre toute la Philosophie Theoretique des Peripateticiens. Ayant parlé de la Metaphysique, il vient à la Physique; Il traicte de toutes les qualitez des Corps, selon diverses opinions, dont les unes sont nouvelles & les autres simplement renouvellées. Il monstre que ce que l'on apelle le Ciel, n'est que l'estenduë de l'Air, & qu'il n'y a que deux Elemens, la Terre & l'Eau; Que le feu n'est point un Element, mais un simple Accident. On ne luy peut accorder entierement ce dernier Article, car si le Feu est estimé un accident, ce n'est qu'à l'esgard de celuy que nous faisons par artifice; Il faut reconnoistre qu'il y en a un autre qui est une veritable Substance, laquelle si elle n'est un Element, doit pourtant estre prise pour un des Corps Principaux qui constituent le Monde."

16 Sorel, *Science des choses corporelles*, 342: "Nous avons donc trouvé que la Terre, l'Eau, l'Air & le Feu sont quatre Corps divers, mais non pas tous quatre Elemens ny Corps simples. La Terre, l'Eau, & l'Air sont estimez simples, mais le Feu est composé. La Terre & l'Eau servent à la composition de tous les Corps, sans que l'Air soit une partie de leur substance lors qu'il se loge quelquefois parmy eux. Le Feu n'estant aussi que ce mesme Air qui est eschauffé, ne doit pas estre du rang des Premiers Corps sinon en tant qu'il agit, non pas qu'il soit partie du composé."

17 *Ibid.*, 343: "Neantmoins à bien considerer toutes choses, il paroist qu'il n'y a que deux sortes de matieres bien distinctes qui soient les Elemens des Elemens, c'est à sçavoir la seche & l'humide. Il est manifeste que la Terre est autre chose que l'Eau, puisque l'Eau rassemble les Atomes de la Terre, & que les petites portions de la Terre qui se sont transmises dans l'Eau, font qu'elle est autre chose que l'Aire. Si la matiere seche estoit semblable à l'humide, elles ne serviroient de rien l'un à l'autre (...). Ce sont les deux qualitez qui constituent les vrais Elemens."

18 *Ibid.*, 329: "[...] il en faut faire l'espreuve, & voir combien il y a de corps simples qui composent les mixtes. Si vous pressez les herbes ou les fleurs & les fruicts des arbres, il en sortira de l'eau, & ce qui

est de terrestre demeurera. Si la chair d'un animal est coupée, le sang & les humeurs en sortent, & le solide en est separé. Voila l'Eau & la Terre; où sont donc l'Air & le Feu?"

19 Sorel, *Perfection de l'homme*, 255-56: "Entre tous les Novateurs on n'en voit point qui s'esloigne d'avantage des Pensées communes. Les Peintures de ses Tourbillons imaginaires (...), ses Figures & quantité d'autres, sont pleines de ces petits Corps si peu connûs, qui y sont representez avec autant d'asseurance que s'il les avoit veûs clairement."

20 Verbeek, *Descartes and the Dutch*, 9.

21 Hereboord, "Consilium de ratione studendi philosophiae," in *Meletemata*, 27-28. It is noteworthy that when he comes to speak of metaphysical authors his students ought to read, Hereboord specifically condemns Vorstius, but later recommends Gorlaeus: "Hoc non praterierim, in Metaphysicis quaestiones discutiendis, speciatim esse insistendum iis, quae moventur in secunda parte speciali de Deo divinisque attributis, ubi fundamenta sunt conquirenda, quibus mens nostra muniri valeat adversus periculosissimos *Conradii Vorstii* in tractatu de Deo errores, quos detexit & solide refutavit *Becanus* in prima parte Theologiae Scholasticae, quae tota in isto genere est optima." He then moves to modern philosophers, who either [i] rejected Aristotle's metaphysics or [ii] proposed their own philosophy. In the former category, he first mentions Ludovicus Vives, Petrus Ramus and Franciscus Patrizzi. "Hisce adjungatur quartus nostri aevi scriptor Belga, Gorlaeus in exercitationibus Philosophicis." The list ends with Campanella, Telesio, Basson, the Boate Brothers (Gerard and Arnold) and Patrizi. As for those who thought up an own, new metaphysics, he cites "illustris Heros, Franciscus Baco de Verulamio," Iohannes Amos Comenius and at the end René Descartes, whom he praises most emphatically.

22 On Ravensperger, see Dibon, *La philosophie néerlandaise*, 211-14; on Ravensperger's astronomical views, see Vermij, *Calvinist Copernicans*, 123-25; for the funerary oration, see Berckringer, *Oratio*.

23 Ravensperger, *Disputatio philosophica inauguralis*, q. 31: "An terra nostra sit magnetica? Affirmant duo saeculi nostri heroes G. Gilbertus ... & Galilaeus de Galilaeis. " I would like to thank Arjen Dijkstra for pointing me to Ravensperger's *Disputatio*.

24 *Ibid.*, q. 4: "An una tantum detur scientia Theoretica?"

25 *Ibid.*, q. 16: "An entis objectivus detur conceptus unus et adaequatus?"

26 *Ibid.*, q. 28: "An ignis & aër sint elementa? Communis & Peripatetica sententia affirmativa est. H. Cardanus *l.2.de.subtil.in pr.& alibi* excludit ignem. Aërem vero etiam Clem. Timplerus *Phys. p. 2.l.3.c.2.q.1.* & D. Gorlaeus *Exer.17.s.1.2.3.4.* Elige, quod voles."

27 *Ibid.*, q. 29: "An aër sua natura nec calidus sit, nec frigidus. In hypothesi, quod elementum non sit, crederem affirmandum cum D. Gorlaeo *l.c.s.5.* & Timpl. *L.c.l.2.c.5.q.5.*"

28 *Ibid.*, q. 36: "An igitur numerus, & unitas materialium sub numero & unitate transcendentali comprehendantur, non secus, ac species sub genere? & numerus ac unitas Arithmetica, seu Quantitativa plane eadem sint cum numero & unitate Transcendentalibus? Aff. [...] in posteriori Avicenna ..., Gorlaeus *exerc.6.princ.* ...."

29 The tradition of viewing Reneri as the first Dutchman known to Descartes exists at least since Burman, *Trajectum eruditum*, 301: "Prima fuit inter Batavos, qui Cartesio innotuit." The tradition of viewing him as the first Cartesian started presumably with Baillet, *Vie de Descartes*, 2: 2.

30 The biographical information on Reneri is taken from Sassen, *Henricus Renerius*; Dibon, *Philosophie néerlandaise*, 197-202; Verbeek, *Descartes and the Dutch*, "Appendix 2: Henricus Reneri," 96-97; Rodis-Lewis, "Descartes," 289; and from personal communications by Robin Buning, who is currently completing a dissertation on Reneri.

31 Bloch, "Pierre Gassendi," 229.

32 Letter by Reneri to Mersenne, probably of March 1638: "Is est mea lux, meus sol, et quod Virgilius in *Bucolicis* dixit, idem possum de ipso dicere: *Erit ille mihi semper Deus...*"; quoted from *Correspondance de Mersenne*, eds. Tannery & De Waard, 7: 115.

33 Verbeek, *Descartes and the Dutch*, 96-97.

34 A further empirical element that is not strictly related with Descartes' interests in those years, are Reneri's microscopical studies of plants and animals, which he described to Mersenne in 1638, saying that he now perceived "what none of the ancients had been able to observe because of their ignorance regarding microscopes." See *Correspondance de Mersenne*, eds. Tannery & De Waard, 7: 115.

35 *Catalogus librorum Reneri*, C3: "Gorlei Exercitationes philos." Gorlaeus is listed right after Gassendi's *Exercitationes* of 1624 and his attack on Fludd of 1630. I owe this reference to Robin Buning.

36 Reneri, *De elementis*, thesis 2: "considerari solent Elementa & corpora ex iis mixtis, sed haec distinctio non est adaequata, 1. quia aërem non comprehendit, ut pote qui nec est elementum nec mixtum ex iis corpus..." I owe the reference to this fascinating disputation to Robin Buning.

37 *Ibid.*, thesis 3: "[...] potest fieri connexio elementorum inter se, ut postea nexus ille fit indissolubilis a causis naturalibus."

38 *Ibid.*, thesis 4: "[...]. Cum nulla phaenomena in elementis sint, quae plus requirant, quam materiam & ejus diversam dispositionem quoad quantitatem, figuram, motum, & quietem." Interestingly, he invokes the Aristotelian Magirus to argue for this (otherwise corpuscular) interpretation of the elements, citing (in thesis 5) the latter's definition of the task of a general (rather than specific) treatment of the elements: "Illa in numero, forma, motu, figura et proprietatibus elementorum est occupata."

39 *Ibid.*, theses 6-9.

40 *Ibid.*, 25: "Sed vis calefaciendi aut etiam urendi in dictis liquoribus aliisve corporibus non adscribenda igni in iis incluso, sed figurae certae, nec non tenuitati ac mobilitati minimarum partium, quibus constant dicta corpora, quibus fiat ut agitatae punctionem igni similem efficiant in linguâ aliove corpore."

41 *Ibid.*, 27: "Aerem autem non constituere partem mixti probabiliter colligitur ex nimia ejus fluiditate ac tenuitate, quae efficiat ut cum crassioribus illis corporibus, terra nimirum & aqua, cohaerere non possit, ut quidem terra & aqua inter se."

42 *Ibid.*, 31: "Aqua & terra in sua puritate considerata, sunt elementa proprie dicta: quia & simplicia sunt, & ad ea mixtorum omnium generationem concurrere arguit vel resolutio quorundum corporum in ea... ."

43 Verbeek, *Descartes and the Dutch*, 97.

44 The biographical information on Regius is taken from Bos, *Correspondence between Descartes and Regius*; Clarke, "Henricus Regius"; Dibon, "Der Cartesianismus in den Niederlanden"; Verbeek, *Descartes and the Dutch*, passim.

45 See Bos, *Correspondence between Descartes and Regius*, 258.

46 Verbeek, *Descartes and the Dutch*, 13; also idem, *Descartes et Regius*, 8; De Vrijer, *Henricus Regius*, 8-20.

47 See Rothschuh, *Physiologie im Wandel.*

48 See Descartes' letters contained in Bos, *Correspondence between Descartes and Regius.*

49 Regius, *Physiologia*, 33 (= *Disputatio de actionibus animalibus, pars prior*, thesis 2); transcribed in Bos, *Correspondence between Descartes and Regius,* 223.

50 Regius, *Physiologia*, 1 (= *Disputationum medicarum primae, De sanitate, pars prior*, thesis 5; transcribed in Bos, *Correspondence between Descartes and Regius*, 199): "Et quamvis ad istas insensibiles particulas alii medici vel philosophi non multum attendere consueverint; nos tamen ex illis innumera naturae mysteria pendere arbitramur."

51 *Ibid.*, 5, thesis 14, transcribed in Bos, *Correspondence between Descartes and Regius*, 202: "Mens, mensura, quies, motus, positura, figura, / Sunt cum materia cunctarum exordia rerum." On the fortuna of this didactic verse, see Bos, "Een kleine geschiedenis."

52 Schoock, *Admiranda methodus*, ch. 5; see Verbeek, *Querelle*, 282.

53 See Rodis-Lewis, "Problèmes discutés entre Descartes et Regius," esp. 36-38.

54 Regius, *Physiologia*, 5-6 (= *Disputationum medicarum primae, De sanitate, pars prior*, thesis 14-15; transcribed in Bos, *Correspondence between Descartes and Regius*, 202): "Idcirco bona temperies a nobis definitur: situs, figura, quantitas, et motus vel quies particularum insensibilium partes sensibiles constituentium, actionibus perficiendis conveniens. A temperie, sive a primis qualitatibus ex quibus constat, omnes aliae corporis humani atque etiam reliquorum omnium tam homogeneorum, quam heterogeneorum corporum qualitates originem ducunt."

55 *Ibid.*, 6, thesis 17 (transcribed in Bos, *Correspondence between Descartes and Regius*, 202): "Calor actualis est varia agitatio insensibilium particularum: frigus autem est earum quies." Gorlaeus, *Exercitationes*, VII.iii.115: "Motu enim corporum crassorum producitur, et attritu nonnunquam."

56 Gorlaeus, *Exercitationes*, VII.iii.115: "Quo modo illud fiat, me latet, & miror illud, sicut plura alia." Descartes, *Principia*, IV.198. This image is discussed in Lüthy, "Where Logical Necessity," 119.

57 See Rodis-Lewis, "Problèmes discutés," 36.

58 See *Testimonium Academiae Ultrajectinae et narratio.*

59 Regius, *De illustribus quaestionibus*, disputation I, thesis 2: "Nullas formas, quae sint substantiae sive completae sive incompletae, in materia, praeter Mentem, esse asserimus. Vera autem rerum naturalium forma est comprehensio motus vel quietis, item situs, figurae et magnitudinis partium, tum insensibilium, tum sensibilium, rebus naturalibus conveniens." I would like to thank Erik-Jan Bos for sharing his copies of these disputations with me.

60 *Ibid.*, disputation II, thesis 4: "Hallucinantur, mea quidem opinione, qui Aristotelis de natura definitionem tanquam omnibus numeris probam defendere student."

61 *Ibid.*, thesis 7: "Materia a magnitudine realiter non differt."

62 *Ibid.*, thesis 11: "Hae non sunt atomi, sed indefinite divisibiles; nec semper ejusdem sunt magnitudinis aut figurae; sed quantum ad talia, idem de ipsis, quod de reliquis corporibus, est dicendum."

63 *Ibid.*, thesis 19: "Nam motus generationis, corruptionis, item accretionis et decretionis sunt tantum varii motus locales particularum insensibilium, qui ad haec tanquam effecta producenda cum particulis materiae concurrunt; ut apparet in generatione vermis ex caseo putrescente, ubi particulae tum insensibiles, tum sensibiles varie disponuntur. Quae dispositio nihil aliud est quam motus localis."

64 *Ibid.*, thesis 22: "Motus (ut primus observavit et docuit Author gallicae dipotricae, horum sacrorum mystagogus) in creatione variis materiae partibus, a Deo varie fuit inditus." I should like to mention that two of the Corollaries do not seem to follow the gist of the Disputation: "I. An quicquid movetur, moveatur sua vi? Aff." [This is not a mechanical understanding of motion]. "II. An sublata subtili materia, detur vacuum? Aff." [This contradicts the identification of extension with matter]. "III. An aer sit siccus? Aff."

65 Beck, *Voetius*, 66.

66 I thank Theo Verbeek for drawing my attention to the difference between the recorded (written) version of the disputation and its oral elaboration.

67 Regius, *De illustribus quaestionibus*, disputation III, thesis 4: "Nos autem omnem motum naturalem esse dicimus: quandoquidem sit secundum naturae leges, nec ullus contra illas fieri potest."

68 *Ibid.*, thesis 7: "Eductione vero formarum substantialium e potentia materiae, quae excogitata est ab iis, qui veras formas ignorarunt, nos non aplius indigere."

69 *Ibid.*, thesis 8: "Forma specialis est mens humana, quia per eam cum forma generali in materia corporea homo est, id quod est. Haec ad formam generalem seu materialem nullo modo potest referri: quoniam ipsa (utpote substantia incorporea) nec est corpus, nec ex motu aut quiete, magnitudine, situ aut figura partium oriri potest."

70 *Ibid.*, thesis 9 and 10: "IX. Ex hac et corpore non fit unum ens per se, sed per accidens, cum singula

sint substantiae perfectae seu completae. X. Cum autem dicuntur incompletae, hoc intelligendum est ratione compositi, quod ex harum unione oritur."

71 See Rodis-Lewis, "Problèmes discutés," 38. Descartes subsequently helped Regius, however, by suggesting a way out of the impasse: he could state (So Descartes proposed in January 1642) that out of the accidental combination of body and soul, a new substance emerged (*ibid.*).

72 Voetius' reasons for worrying about Descartes' philosophy are excellently analyzed in Van Ruler, *Crisis of Causality*, notably ch. 1; see also, Goudriaan, *Reformed Orthodoxy*, esp. 234-42 and *passim*.

73 See Verbeek, *Querelle*, 30.

74 *Ibid.*, 44.

75 Descartes' viewpoint is expressed in his "Letter to Dinet" (*Oeuvres*, 7: 563-603), which also summarizes Voetius' historical pedigree of the view that man is an accidental being; see Verbeek, *Querelle*, 143.

76 *Testimonium*, 28. The Latin original is given above, in chapter 3, footnote 169. See also Verbeek, *Querelle*, 98. Note that this historically rich version is the draft version of the corollary. At the public disputation of 18 December, a shorter version was defended.

77 *Testimonium*, 29: "Negative Philosophia de substantialibus rerum formis earumque propriis ac specificis facultatibus, seu qualitatibus activis, et consequenter specificis rerum et distinctis naturis (quam Taurellus, Gorlaeus, et Bassonis in scenam nostra hac aetate reducere conati) non satis cum Physica Mosaica et sacra hactenus conciliari posse videtur. Consulant studiosi nostri Danaeum, Zanchium, Commentatores in Genesin; &c. scholasticos ad Lombardum et Thomam."

78 See Verbeek, "'Ens per accidens"; Bos, *Correspondence between Descartes and Regius*. Rodis-Lewis, 39, asks whether Regius, in his 1642 rebuttals, abandoned Gorlaeus, "son premier inspirateur?" She answers: "Tout en répétant les précisions de Descartes sur l'unique forme substantielle (alors que Gorlaeus les niait toutes), il avançait incidemment que l'union qui fait l'homme, sans essence propre "esse tantum entis modum," tout en distinguant cet être modal d'un être de raison."

79 Letter of Descartes to Mersenne, 8 October 1629: "Pour la raréfaction, je suis d'accord avec ce medecin [Basson], et ay maintenant pris party touchant tous les fondements de la philosophie; mais peut-estre je n'explique pas l'aether comme luy." Beeckman, *Journal*, fol. 177bis verso, writes in 1623: "Philosophia naturalis Sebastiani Bassonis incidens in manus meas, visa adhuc est parum aut nihil alienum ab iis, quae in hoc libro explicuimus, tractare." On Basson's Life, Doctrine and Influence, see Lüthy, "Sébastien Basson."

80 On Voetius' view of Basson and Gorlaeus, see Van Ruler, *Crisis of Causality*, ch. 7.

81 Descartes, "Lettre à Dinet," in *Oeuvres*, 7: 563-603, at 586.

82 Note that the editors of the *Correspondance de Marin Mersenne*, 1: 491-92, also point to Bruno, Hill, Bacon, Basson, Beeckman and Gorlaeus for anticipating Descartes' idea (first adopted in the *Regulae*) that a subtle corpuscular fluid – aether, fire or spirit – filled the interstitial voids of bodies.

83 Hattab, *Descartes on Forms*, 172 and 219.

84 Pasnau, *Metaphysical Themes*, 599.

85 The thesis is reproduced in the *Testimonium Academiae Ultrajectinae*; see Verbeek, *Querelle*, 109, for a translation.

86 *Testimonium Academiae Ultrajectinae*; see Verbeek, *Querelle*, 112-13, for a translation.

87 Schoock, *Admiranda methodus*, Preface, s.p. [p. xxxiv]: "[...] Gorlaeum et Taurellum [...] authores in vulgus notissimos [...]. [...] An vero Voetius ignoret Gorlaeum ac Taurellum discat ex studiosis, qui authores eos, quotidie ab eo commodato accipere solent, aut si iis fidem deroget in consilium adhibeat suum Medicum." This is a reply to Descartes' allegation that Voetius was not acquainted with these two authors.

88 *Ibid.*, 6: "[...] sententiae antiquorum Philosophorum interpolator ineptus Basso, per disputandi pruriginem Atheismo proximus Nicol. Taurellus, et ejus sectator quamvis haut aequis passibus, Gorlaeus?"

89 *Ibid.*, 42: "[...] ne quid dicam de omnibus, quae in orbe Europaeo sunt Acadademiae et Scholae, quas certo scio ad unam omnes Cartesii novam philosophiam carbone notaturas, unoque plebiscito cum Taurelli, Gorlaei, ac Bassonis deliriis proscripturas esse."

90 See Voetius, *Selectae disputationes theologicae*, 120-29.

91 On Voetius' reactions to contemporary developments in philosophy and physics, see Verbeek, *Descartes and the Dutch*, ch. 1 ("Prologue") and ch. 2 ("The Utrecht Crisis"); Ruler, *Crisis of Causality*, ch. 1 and *passim*.

92 Van Berkel, *Isaac Beeckman en de mechanisering*, notably chapters 7 and 8.2. Cf. also De Buzon, "Beeckman, Descartes."

93 Van Berkel, *Mechanical Philosophy*, [forthcoming], ch. 1, n. 39. Incidentally, Jacob Beeckman enrolled at the same time at Franeker as Vorstius' Steinfurt students Welsing and Omphalius, who were the driving forces behind the publication of the scandalous Socinian treatise; see Fockema Andrae e.a., *Album studiosorum*, 47.

94 Beeckman, *Journal*, 1: 51.

95 On Beeckman's debt to ancient atomism, see Gemelli, *Isaac Beeckman, atomista*.

96 Beeckman, *Journal*, 1: 244.

97 The famous picture of the metaphysical roots and the physical trunk are found in the Preface to the French edition of the *Principia*. One interesting reconstruction of Descartes' perceived need to deduce his natural philosophy from a metaphysical set of principles is found in Henry, "Metaphysics and the Origins."

98 Hattab, *Descartes on Forms*, 159, and chapter 7 ("Atoms, modes, and other heresies").

99 Bayle, *Dictionnaire*, s.v. "Gorlaeus," 160-61:" Ens per se, Ens per accidens, sont des phrases inexplicables, un vrai jargon des logiciens espagnols, qui ne signifie rien [...]."

100 The demise of Voetius' world can for example be assessed by the fact that one of his most faithful followers, and also one of his last ones, Gerard de Vries, edited the third edition of Gassendi's *Disquisitio metaphysica* in 1691. See Verbeek, "Gassendi et les Pays Bas," 263.

101 Verhel, *Speculum philosophiae primae entis*, dedication; see above, p. 15.

102 Daniel, *Voyage du monde de Descartes*, 184-85, quoted from Roux, "An Empire Divided."

103 Feuerlein, *Taurellus defensus*, xv-xvi, casts Gorlaeus in the opposite role; here, he appears as a footnote to Taurellus.

104 Menn, "Intellectual Setting," 34.

105 Morhof, *Polyhistor*, pars II, lib. II, cap. 1, sect. 3, p. 273: "Laudem certe meretur, quod ante Cartesium ista videret, quae postea Cartesius dogmata sua esse voluit."

106 Reimmann, *Versuch einer Einleitung*, 3: 563: "Denn alle diese Einwürfe [*sc.* von Voetius] haben die Cartesianer, welche die meisten Hypotheses Gorlaeanas hernachmals in ihr Systema philosphorum aufgenommen, dem Voëtio beantwortet."

107 See, e.g., Goudriaan, *Philosophische Gotteserkenntnis*, "Einleitung."

108 Leinsle, *Das Ding und die Methode*, 147-65; 164. It would be worth studying the similarities and dissimilarities between the definition of *ens* in Suárez, Taurellus and Gorlaeus; see Darge, "Grundlegung."

109 Hattab, *Descartes on Forms*, ch. 7, repeatedly claims that Gorlaeus' philosophy betrays his reading of Suárez' metaphysics. While it seems to me that Gorlaeus could have taken what Hattab calls "Suárez' innovations" (*ibid.*, 180) instead from Taurellus (whom Hattab does not examine), I admit that this question deserves further investigation.

110 Kuhn, *Essential Tension*, xi-xiii.

111 Hunt, *Chemical and Geological Essays*, 428, 450.

112 William R. Newman has documented the chemical prehistory of atomism in numerous publications. For a good synthesis of his main thesis, see his *Atoms and Alchemy*.

113 See *Testimonium* in Verbeek, *Querelle*, 111.

114 On the lack of validity of the alleged empirical argument in favor of atomism, see Kangro, "Erklärungswert"; Meinel, "Early Seventeenth-Century Atomism"; idem, "'Das letzte Blatt'."

115 For a comparison of Bruno's and Taurellus' atomism, see Lüthy, "Entia & sphaera."

116 See Van Ruler, *Crisis of Causality*. On the "confessionalization of physics," see Leijenhorst & Lüthy, "Erosion of Aristotelianism."

Bibliography

Note that the alphabetical order takes into account prefixes such as "van," "de," "von," etc., to avoid arbitrariness. Also, publications by any given author are ordered alphabetically, not according to their year of publication.

MANUSCRIPTS

Basel

Universitätsbibliothek, MS Frey-Gryn. MS. II.4. no. 307: Letter of Nicholaus Taurellus to Theodor Zwinger, 19 October 1580.

Universitätsbibliothek, MS. A II 6: *Matricula facultatis theologicae,* 1597 - .

Leeuwarden

Tresoar Leeuwarden, Auction Catalogue Auletius, MS. 13-13, inv.nr. 191, fol. 28r-35v, cont. 22r-26v, edited by M.H.H. Engels, http://mpaginae.atspace.com/Auletiustxtimg.htm.

Hystoarysk Sintrum Ljouwert, Toegang 178, Martena e.o. MR 93: Testament of David van Goorle.

Leiden

Universiteitsbibliotheek, MS Papenbroeck N. 21: Engelbert Aegidii & Engelbert van Engelen, *Album Amicorum.*

Archive of Senate and Faculty, *Album studiosorum 1575-1516*, MS. ASF 7.

Regional Archive, Oud Notarieel Archief 210, nr. 174 (31 August 1624): Testament of Jacob Commelin.

Vatican City

Archivio della Congregatione per la Dottrina della Fede, Decreta Sancti Officii, vols 1597, 1598: Inquisitorial acts with respect to Henricus de Veno.

PRINTED SOURCES

Abels, Paul, "Das Arnoldinum und die Niederlande während seiner ersten Blütezeit: Das Verhältnis einer Hassliebe," in *400 Jahre Arnoldinum 1588-1988*, ed. Kreisheimatbund Steinfurt (Greven, 1988), 78-97.

Adama, Lollius, *Fortunati Crellii Isagoges logicae, in duas partes tributae, in communem et in propriam* (Franeker, 1609).

——, *Theses logicae, de natura logices.* [Resp.] Liebwertus a Wynshem (Franeker, 1606).

Adresboek Nederlandse drukkers en boekverkopers tot 1700, eds. J.A. Gruys and Jan Bos for STCN (The Hague, 1999).

Album studiosorum Academiae Lugduno Batavae MDLXX-MDCCCLXXV, ed. Willem Nicolaas Du Rieu (The Hague, 1875).

Allgemeine Deutsche Biographie, ed. Königliche Akademie der Wissenschaften (Leipzig, 1875).

Andreas, Valerius, *Bibliotheca Belgica*, rev. edition (Leuven, 1643).

Ariew, Roger, *Descartes and the Last Scholastics* (Ithaca, NY, 1999).

Arminius, Jacob, *Opera theologica* (Leiden, 1629).

——, *Verclaringhe Iacobi Arminii saliger ghedachten, in zijn leven professor theologiae binnen Leyden* (Leiden, 1610).

Augustijn, Cornelis, "Het Nederlandse Protestantisme in vogelvlucht," *Reformatica. Teksten uit de geschiedenis van het Nederlandse protestantisme* (Zoetermeer, 1996), 25-43.

Bacon, Francis, *Instauratio magna* (London, 1620).

——, *On the Proficience and Advancement of Learning, Divine and Human* (London, 1605).

Baillet, Adrien, *La Vie de M. Des Cartes*, 2 vols. (Paris, 1691).

Bakhuizen van den Brink, Jan Nicolaas, ed., *De Belijdensgeschriften volgens artikel X van de Kerkorde van de Nederlandse Hervormde Kerk* (The Hague, 1966).

Baffioni, Carmela, *Il quarto libro dei "Meteorologica" di Aristotele* (Naples, 1981).

Baldini, Ugo & Leen Spruit, eds., *Catholic Church and Modern Science. Documents from the Historical Archives of the Roman Holy Office and the Index.* Vol. 1: *Sixteenth-Century Documents*, 4 vols. (Rome, 2009).

Basson, Sébastien, *Philosophiae naturalis adversus Aristotelem libri XII. In quibus abstrusa veterum physiologia restauratur, & Aristotelis errores solidis rationibus refelluntur* (Geneva, 1621).

Baudartius, Willem, *Memoryen ofte cort verhael der gedenck-weerdichtse so kercklicke als werltlicke gheschiedenissen*, 2 vols. (2nd ed., Zutphen, 1624).

Bayle, Pierre, *Dictionnaire historique et critique*, 4 vols., 5th ed. (Amsterdam, 1740; repr. Geneva, 1995).

Beck, Andreas Johannes, *Gisbertus Voetius (1589-1676). Sein Theologieverständnis und seine Gotteslehre* (Göttingen, 2007).
Bedenckingen over de Beroepinghe D.D. Conradi Vorstii, tot de Professie der H. Theologie in de Universiteit tot Leyden... (S.l., 1611).
Beeckman, Isaac, *Journal tenu par Isaac Beeckman de 1604 à 1634*, ed. Cornelis de Waard, 4 vols. (The Hague, 1939-1953).
Berckringer, Daniel, *Oratio in obitum Jacobi Ravensbergeri* (Utrecht, 1650).
Bergsma, Wiebe, *Tussen Gideonsbende en publieke kerk. Een studie over het gereformeerd protestantisme in Friesland, 1580-1650* (Hilversum & Leeuwarden, 1999).
Bertius, Petrus, *Aenspraeck aan D. Fr. Gomarum op zijne Bedenkinghe over de Lijckoratie ghedaen na de begrafenisse van D. Jacobus Arminius zaliger* (Leiden, 1609).
——, *De elementis ...* [resp.] Thodorus Oortius (Leiden, 1599).
——, *De mixtione ...* [resp.] Samuel Nothaeus (Leiden, 1604).
——, *Liickoratie over den doot vanden Eerwaardighen ende wytberoemden Heere Jacob Arminius* (Leiden, 1609).
Bibliotheca Belgica. Bibliographie générale des Pays-Bas, ed. Ferdinand van der Haeghen et al., republished by Marie-Thérèse Lenger (Brussels, 1964).
Biografisch lexicon voor de geschiedenis van het Nederlands protestantisme, eds. Doede Nauta & Johannes van den Berg, 3 vols. (Kampen, 1978-2006).
Biographisch Woordenboek der Nederlanden, bevattende levensbeschrijvingen van zoodanige personen, die zich op eenigerlei wijze in ons vaderland hebben vermaard gemaakt, ed. Abraham Jacob van der Aa, 12 vols. (Haarlem, 1852-1878).
Blair, Ann, "Mosaic Physics and the Search for a Pious Natural Philosophy in the Late Renaissance," *Isis*, 91 (2000), 32-58.
Bloch, Olivier, "Pierre Gassendi," in Jean-Pierre Schobinger, ed., *Überweg Grundriss der Geschichte der Philosophie: Die Philosophie des 17. Jahrhunderts*, 4 parts in 8 vols. (Basel, 1998-2001), 2/1: 201-32.
Blockmans, Willem Pieter, "The Formation of a Political Union, 1300-1600," in Blom & Lamberts, eds., *History of the Low Countries*, 55-140.
Blom, Johan Cornelis Hendrik & Emiel Lamberts eds., *History of the Low Countries*, transl. James C. Kennedy (New York, 1999).
Boeles, Willem Boele Sophius, *Frieslands Hoogeschool en het Rijks Athenaeum te Franeker*, 2 pts in 3 vols. (Leeuwarden, 1879).
Bögels, Theo S.J.G., *Govert Basson. Printer, Bookseller, Publisher: Leiden, 1612-1620* (Nieuwkoop, 1992).
Bos, Erik-Jan, *The Correspondence between Descartes and Henricus Regius* (Utrecht, 2002).
——, "Een kleine geschiedenis van een cartesiaans versje," in M. van Egmond, B. Jaski & H. Mulder, eds., *Bijzonder onderzoek: Een ontdekkingsreis door de Bijzondere Collecties van de Universiteitsbibliotheek Utrecht* (Utrecht, 2009), 244-51.

Bosch, L.J.M., *Petrus Bertius 1565-1629* (Ph.D. thesis, University of Nijmegen, 1979).

Bots, Hans (J.A.H.G.M.) & Wilhelm Th.M. Frijhoff, "De studentenpopulatie van de Franeker academie: een kwantitatief onderzoek (1585-1811)," in Jensma et al., eds., *Universiteit te Franeker*, 56-72.

Brahe, Tycho, *Opera omnia*, ed. John L.E. Dryer, 15 vols. (Copenhagen, 1913-1929).

Brandt, Geeraert, *Historie der Reformatie, en andre kerkelyke geschiedenissen, in en ontrent de Nederlanden*, 4 vols. (Amsterdam and Rotterdam, 1671-1704).

Briels, Johannes Gerardus Carolus Antonius, *Zuidnederlandse boekdrukkers en boekverkopers in de Republiek der Verenigde Nederlanden omstreeks 1570-1630: een bijdrage tot de kennis van de geschiedenis van het boek* (Nieuwkoop, 1974).

Bruno, Giordano, *De triplici minimo et mensura* (Frankfurt, 1591).

Buccaferrea, Ludovico, *Lectiones in quartum Meteororum Aristotelis librum* (Venice, 1563).

Burger, C.P., Jr., "De boekverkopers Commelin te Genève, Heidelberg en Leiden," *Tijdschrift voor boek- en bibliotheekswezen* 9 (1911), 145-76.

Burmannus, Caspar, *Traiectum eruditum, virorum doctrina inlustrium, qui in urbe Trajecto ... nati sunt ... vitas, fata et scripta* (Utrecht, 1738).

Cardano, Girolamo, *De subtilitate libri XXII* (Nuremberg, 1550).

——, *In calumniatorem librorum de subtilitate, actio prima*, in: *De subtilitate libri XXI*. Third rev. ed., (Basel, 1560), 1265-1426.

——, *Contradicentium medicorum libri II* (Paris, 1565).

Casmann, Otto, *Angelographia, sive Commentationum disceptationumque physicarum prodromus problematicus* (Frankfurt, 1605).

——, *Cosmopoeia et* Ουρανογραφία *Christiana* (Frankfurt, 1598).

Catalogus variorum ac rarissimorum librorum in omnibus scientiis & facultatibus, clarissimi, doctissimique viri D. Henrici Reneri... (Utrecht, 1639).

Catechesis religionis Christianae, quae traditur in ecclesiis et scholis Palatinatus (Heidelberg, 1563).

Christianson, John Robert, *On Tycho's Island. Tycho Brahe and His Assistants, 1570-1601* (Cambridge, 2000).

Clarke, Desmond, "Henricus Regius," *The Stanford Encyclopedia of Philosophy (Summer 2011 Edition)*, ed. Edward N. Zalta (URL = <http://plato.stanford.edu/archives/sum2011/entries/henricus-regius/>)

Clericuzio, Antonio, *Elements, Principles and Corpuscles. A Study of Atomism and Chemistry in the Seventeenth Century* (Dordrecht, 2000).

Correspondance du P. Marin Mersenne, religieux minime, eds. Marie Tannery-Prisset, Cornelis de Waard, Jr. & René Pintard, 17 vols. (Paris, 1932-1988).

Corvinus, Johannes Arnoldi, *Christelicke ende ernstighe vermaninghe tot vrede aen R. Donteclock*... (The Hague, 1609).

Cossee, Eric H. "Arminius and Rome," in Van Leeuwen et al, eds., *Arminius, Arminianism, and Europe*, 73-85.
Cudworth, Ralph, *The True Intellectual System of the World* (London, 1678).
Danaeus, Lambertus [= Lambert Daneau], *Physica christiana, sive de rerum creatarum cognitione et usu* (Leiden, 1576).
——, *Physices Christianae pars altera, sive de rerum creatarum natura* (Geneva, 1580)
Daniel, Gabriel, *Voyage du monde de Descartes* (Paris, 1691).
Darge, Rudolf, "Die Grundlegung einer allgemeinen Theorie der transzendentalen Eigenschaften des Seienden bei F. Suarez," *Zeitschrift für philosophische Forschung*, 54 (2000), 341-64.
Daugirdas, Kestutis, "The Biblical Hermeneutics of Socinians and Remonstrants in the Seventeenth Century," in Van Leeuwen et al, eds., *Arminius, Arminianism, and Europe*, 89-113.
De Buzon, Frédéric, "Beeckman, Descartes and Physico-Mathematics," in *The Mechanization of Natural Philosophy*, eds. Daniel Garber and Sophie Roux (Berlin, 2012), [forthcoming].
De Crane, J.W., "Het aloud geslacht *Martena*, in genealogisch overzigt voorgesteld," *De vrije Fries*, 1.3 (1838), 189-232.
Den Arminiaenschen Dreck-waghen Gheheel Naer het leven afghebeelt ghelijck de Letters binnen uytwissen sullen (Amsterdam, 1618).
Den Boer, William, "Jacobus Arminius: Theologian of God's Twofold Love," in Van Leeuwen et al, eds., *Arminius, Arminianism, and Europe*, 25-50.
De Ridder-Symoens, Hilde, "Buitenlandse studenten aan de Franeker Universiteit 1585-1811," in Jensma et al., eds, *Universiteit te Franeker*, 73-89.
Descartes, René, Œuvres, ed. Charles Adam & Paul Tannery, 11 vols. (Paris, 1964-1974).
——, *Principia philosophiae* (Amsterdam, 1644).
Des Chene, Dennis, "Cartesiomania," *Perspectives on Science* 3 (1995), 534-58.
d'Espagnet, Jean, *Enchyridion physicae restitutae...Tractatus alter inscriptus: Arcanum hermeticae philosophiae* (Paris, 1623).
——, *La Philosophie naturelle restablie en sa pureté*, ed. and transl. Jean Bachout (Paris, 1651).
De Veno, Henricus, *Disputationum physicarum prima, de physica*. [Resp.] Bernardus Paludanus (Franeker, 1603). [London, The British Library, 7306 f.6 no. 38].
——, *Disputationum physicarum secunda, de principiis et causis rerum naturalium*. [Resp.] Duca a Buvama (Franeker, 1604). [Universitätsbibliothek Erlangen-Nürnberg].
——, *Disputationum physicarum quarta, de infinito et loco*. [Resp.] Pierius H. Winsemius (Franeker, 1604). [Geneva, Bibliothèque Publique et Universitaire, Cd. 145-48; Paris, Bibliothèque Nationale, R. 2102].

——, *Disputationum physicarum quinta, de mundo in genere.* [Resp.] Gualtherus Zellius (Franeker, 1604). [Geneva, Bibliothèque Publique et Universitaire, Cd. 145-41].
——, *Disputationum physicarum octava, de elementis.* [Resp.] Duca a Buvama (Franeker, 1604). [Geneva, Bibliothèque Publique et Universitaire, Cd. 145-61].
——, *Disputationum physicarum nona, de misti generatione et ejus interitu.* [Resp.] Bernardus Paludanus (Franeker, 1604). [London, British Library, 7306 f.6 no. 46].
——, *Dissertatio politica de magistratu.* [Resp.] Hero a Weerdum et Inhusen (Franeker, 1606). [Leeuwarden, Tresoar, A 1669].
——, *Quaestiones illustres ex tota philosophia conquisitae, quas ad summum in philosophiae gradum consequendum publice ventilandas proponit Hieronimus Fredericus Isbrandi* (Franeker, 1605). [Leeuwarden, Tresoar, A 1669].
——, *Syllabus errorum et contradictionum I[usti] Lipsii. Quae in Polit[ica]* (s.l., s.d., Franeker [?], 1604[?]) [New York Public Library, SEB p.v. 20, no. 3].
——, *Theoremata physica de aëre.* [Resp.] Petrus Theocari (Franeker, 1604). [Paris, Bibliothèque Nationale, R. 2217]
——, *Theorema physica de anima rationali et eius facultatibus.* [Resp.] Petrus Theocari (Franeker, 1604). [Geneva, Bibliothèque Publique et Universitaire, Cd. 145-97].
De Vrijer, Marinus J.A., *Henricus Regius. Een 'cartesiaansch' hoogleraar aan de Utrechtsche Hoogeschool* (The Hague, 1917).
Dibon, Paul, *La philosophie néerlandaise au siècle d'or.* Vol. 1: *L'enseignement philosophique dans les universités à l'époque précartésienne (1575-1650)* (Paris, 1954).
——, "Der Cartesianismus in den Niederlanden," in Jean-Pierre Schobinger, ed., *Überweg Grundriss der Geschichte der Philosophie: Die Philosophie des 17. Jahrhunderts*, 4 parts in 8 vols. (Basel, 1998-2001), 2/1: 349-74.
Dijstelberge, Paul, *De beer is los! Ursicula: een database van typografisch materiaal uit het eerste kwart van de zeventiende eeuw als instrument voor het identificeren van drukken* (PhD thesis, Amsterdam University, 2007).
Dilthey, Wilhelm, *Das natürliche System der Geisteswissenschaften im 17. Jahrhundert* [1892-1893], in Dilthey, *Gesammelte Schriften*, vol. 2: *Weltanschauung und Analyse des Menschen seit Renaissance und Reformation*, ed. Georg Misch, 7th ed. (Stuttgart, 1964).
Düring, Ingemar, *Aristotle's Chemical Tratise Meteorologica, Book IV, with Introduction and Commentary* (Göteborg, 1944).
Eglisemmius, Georg, *Crisis Vorstiani responsi. Qua D. Conrad Vorstius denuo Atheismi, Ethnicismi, Judaismi, Turcismi, Haereseos, Schismatis, et ignorantiae arguitur* (Delft, 1612).
Ekkart, Rudolf Erik Otto, *Franeker professorenportretten: iconografie van de profes-*

soren aan de Academie en het Rijksathenaeum te Franeker 1585-1843 (Franeker, 1977).
Emerton, Norma E., *The Scientific Reintrpretation of Form* (Ithaca, NY, 1984).
Engels, M.H.H., ed., *Brieven aan de Curator van de Universiteit van Franeker Johannes Saeckma (1572-1636). Uitgave van de Codex Saeckma en de brieven van 12 overeenkomstige correspondenten uit de verzameling Gabbema*, 2 vols. (Leeuwarden, 1995).
Faber, Timaeus, "Disputatio sexta, ad tit. 8.9.10... respondente Frederico Stellingwerf," in *Disputationes anniversariae ad IIII libros institutionum imperialium, in illustri Frisorum Academia Franecarana propositae a Timaeo Fabro IC* (Franeker, 1609).
Fechner, Gustav Theodor, *Über die physikalische und philosophische Atomenlehre*, 2nd rev. ed. (Leipzig, 1864; Frankufrt, 1982).
Fernel, Jean, *De abditis rerum causis* (Paris, 1548).
——, *Jean Fernel's* On the Hidden Causes of Things. *Forms, Souls and Occult Diseases in Renaissance Medicine*, transl. by John M. Forrester, intro. by John M. Forrester & John Henry (Leiden, 2005).
Feuerlin, Jacob Wilhelm, *Taurellus defensus* (Nuremberg, 1734).
Firpo, Luigi, *Il processo di Giordano Bruno*, ed. D. Quaglioni (Rome, 1993).
Fockema Andrea, Sybrandus Johannes & Theodorus Jospenus Meijer, *Album studiosorum Academiae Franekerensis (1585-1811, 1816-1844)*. Vol. 1: *Naamlijst der studenten* (Franeker, 1968).
Foppens, Jean-François, *Bibliotheca Belgica, sive virorum in Belgio vita, scriptisque illustrium catalogus, librorumque nomenclatura: continens scriptores a clarissimis viris Valerio Andrea, Auberto Miræo, Francisco Sweertio, aliisque, recensitos, usque ad annum M.D.C.LXXX.* (Brussels, 1739).
Foxius Morzillo, Sebastiano, *Ethices philosophiae compendium, ex Platone, Aristoteli aliisque optimis quibus auctoribus collectum* (Heidelberg, 1561).
Freedman, Joseph S., *European Academic Philosophy in the Late Sixteenth and Early Seventeenth Centuries: The Life, Significance, and Philosophy of Clemens Timpler (1563/64-1624)*, 2 vols. (Hildesheim, 1988).
——, "The Soul (*anima*) according to Clemens Timpler (1563/4-1624) and Some of this Central European Contemporaries," in Barbara Mahlmann-Bauer, ed., *Scientia et Artes. Die Vermittlung alten und neuen Wissens in Literatur, Kunst und Musik* (Wolfenbüttel, 2004), 792-830.
Fruin, Robert Jacobus, *Tien Jaren van den Tachtigjarigen Oorlog* (Den Haag, 1904).
Fuller, Thomas, *The Church History of Britain from the Birth of Jesus Christ, until the Year 1648* (London, 1655).
Galama, Sybrand Haije Michiel, *Het wijsgerig onderwijs aan de Hogeschool te Franeker 1585-1811* (Franeker, 1954).

Garber, Daniel, *Descartes' Metaphysical Physics* (Chicago, 1992).

Gassendi, Pierre, *Exercitationum paradoxicarum adversus Aristoteleos libri septem, in quibus praecipua totius Peripateticae doctrinae atque dialecticae excutiuntur; opiniones vero aut novae, aut ex vetustioribus obsoletae stabiliuntur, liber primus: In doctrinam Aristoteleorum universe* (Grenoble, 1624; 2nd ed., Amsterdam, 1649)

——, *Syntagma philosophiae Epicuri cum refutationibus dogmatum quae contra fidem christianam ab eo asserta sunt* (Lyon, 1649).

Gaudenzio, Paganino, *De Pythagoræa animarum transmigratione opsculum. Accecdunt ... De aperipato Caesaris Scaligeri exercitationes* (Pisa, 1641).

Gaudriaan, Aza, *Philosophische Gotteserkenntnis bei Suárez und Descartes im Zusammenhang mit der niederländischen reformierten Theologie und Philosophie des 17. Jahrhunderts* (Leiden, 1999).

Galilei, Galileo, *Il Saggiatore, nel quale con bilancia esquisita e giusta si ponderano le cose contenute nella libra astronomica e filosofica di Lotario Sarsi...* (Rome, 1623).

Gemelli, Benedino, *Isaac Beeckman. Atomista e lettore critico di Lucrezio* (Florence, 2002).

Goclenius, Rudolph, *Problemata logica*, 5 vols. (Marburg, 1596-99).

Gomarus, Franciscus, *Bedencken over de lijck-oratie van meester P. Bertius* (Leiden, 1609).

Gorlaeus, Abraham, *Dactyliotheca seu Annulorum sigillarium quorum apud Priscos tam Graecos quam Romanos usus. E ferro aere argento & auro promptuarium. Acc. variarum gemmarum ... scalpturae* (s.l., s.a. [Leiden and Delft, 1601]).

——, *Thesaurus numismatum Romanorum aureorum argenteorum aereorum ad familias eius urbis: usque ad obitum Augusti* (Antwerpen, 1605, Leiden and Delft, 1608, 1609).

——, *Thesaurus numismatum Romanorum, ad familias ejus urbis spectantium, usque ad obitum Angusti. Acc. ejusdem Paralipomena seu typi numorum Roman. usque a Fulvio Ursino non editos idem posside* (Delft, 1609).

Gorlaeus, David, *Exercitationes philosophicae post mortem auctoris editae, quibus universa fere discutitur philosophia theoretica, et plurima ac praecipua Peripateticorum dogmata evertuntur* (Leiden, 1620).

——, *Idea physica, cui adjuncta est Epistola cuiusdam Anonymi de terrae motu* (Utrecht, 1651). The Latin text is copied and translated into Dutch and Frisian in *Davidis Gorlaei Ultrajectini, Idea Physicae (1651), met vertalingen in het Nederlands en Fries*, edited and intoduced by F.A.H. Peeters (Tilburg, 1986).

Goudriaan, Aza, "'Augustinus Asleep' or 'Augustinus Awake'? Jacobus Arminius' Reception of Agustine," in Van Leeuwen et al, eds., *Arminius, Arminianism, and Europe*, 51-72.

——, *Reformed Orthodoxy and Philosophy, 1625-1750. Gisbertus Veotius, Petrus van Mastricht, and Anthonius Driessen* (Leiden, 2006).

Gregory, Tullio, "Studi sull'atomismo del Seicento. II. David van Goorle e Daniel Sennert," *Giornale critico della filosofia italiana*, 45 (1966), 44-63.

Groot, Aart de, "Franeker als Irenopolis. F. Socinus, « De officio hominis christiani », 1610," in W. Otten & W.J. van Asselt, eds., *Kerk en conflict. Identiteitskwesties in de geschiedenis van het Christendom* (Zoetermeer, 2002), 102-14.

Gruys, Johannes A., "Een Leidse veilingscatalogus te Brussel, of: De Amsterdamse Commelins te Leiden," in C. Coppens et al, eds., *E codicibus impressisque. Opstellen over het boek in de Lage Landen voor Elly Cockx-Indestegen* (Leuven, 2004), III: 393-417.

——, "Ornamental Bears and Other Animals. An Excursion on Some Head and Tail Pieces," in R. Breugelmans, ed., *The Pilgrim Press* (Nieuwkoop, 1987), 161-69.

Gualtherus, Marcus, *De vita et obitu reverendi, clarissimi et doctissimi viri, Dn. Conradii Vorstii SS. Theologiae Doctoris* (s.l., 1624).

Güldner, Gerhard, *Das Toleranz-Problem in den Niederlanden im Ausgang des 16. Jahrhunderts* (Lübeck, 1968).

Gysens, Steven, "Vossius, Gerard," in *Nationaal Biographische Woordenboek* (Brussels, 1992), 14: 734-40.

——, "Gerardus Vossius, éditeur et traducteur de textes patristiques. L'exemple de la *Vita et encomium Gregorii Thaumaturgi* par Greegoire de Nysse," in *Philohistor: Miscellanea in honorem Caroli Lega septuagenarii*, ed. A. Schoors & P. Van Deun (Leuven, 1994), 553-68.

Hamerster, Dominicus, *Nauwkeurige en duidelijke verklaring over de statuten, ordonnantien, reglementen en costumen van rechte in Friesland, part 1* (Leeuwarden, 1743).

Hannequin, Arthur, *Essai critique sur l'hypothèse des atomes dans la science contemporaine*, 2nd ed. (Paris, 1899).

Harnack, Adolf (von), *Lehrbuch der Dogmengeschichte.* Vol. III: *Die Entwicklung des kirchlichen Dogmas* (Freiburg, 1897).

Harrison, Archibald Harold Walter, *The Beginnings of Arminianism to the Synod of Dort* (London, 1926).

Hattab, Helen, *Descartes on Forms and Mechanisms* (Cambridge, 2009).

Heereboord, Adriaan, *Meletemata philosophica, maximam partem, metaphysica* (Leiden, 1654).

Henry, John, "Atomism and Eschatology: Catholicism and Natural Philosophy in the Interregnum," *British Journal for the History of Science*, 15 (1982), 211-39.

——, "A Cambridge Platonist's Materialism: Henry More and the Concept of Soul," *Journal of the Warburg and Courtauld Institutes*, 49 (1986), 172-95.

——, "Occult Qualities and the Experimental Philosophy: Active Principles in pre-Newtonian Matter Theory," *History of Science*, 24 (1986), 335-81.

——, "Henry More versus Robert Boyle: The Spirit of Nature and the Nature of

Providence," in Sarah Hutton, ed., *Henry More (1614-1687): Tercentenary Studies* (Dordrecht, 1990), 55-75.
——, "Henry More and Newton's Gravity," *History of Science*, 31 (1993), 83-97.
——, "Metaphysics and the Origins of Modern Science: Descartes and the Importance of Laws of Nature," *Early Science and Medicine*, 9 (2004), 73-114.
Heuermann, Georg, *Geschichte des reformierten gräflich Bentheimschen Gymnasium Illustre Arnoldinum zu Burgsteinfurt* (Burgsteinfurt, 1878).
Hill, Nicholas, *De philosophia epicurea, democritica, theophrastica, proposita simpliciter, non edocta* (Paris, 1601; 2nd ed. Geneva, 1619).
Hirai, Hiro, *Le concept de semence dans les théories de la matière à la Renaissance: de Marsile Ficin à Pierre Gassendi* (Turnhout, 2005).
Holtzapffel, Koen & Marius van Leeuwen, eds., *De Remonstrantie 400 jaar. Ontstaan, historie, actualiteit* (Zoetermeer, 2010).
Hooykaas, Reyer, *Geschiedenis der natuurwetenschappen van Babel tot Bohr* (Utrecht, 1971).
——, "The Experimental Origin of Chemical and Molecular Theory before Boyle," *Chymia* 2 (1949), 65-80.
——, *G.J. Rheticus' Treatise on Holy Scripture and the Motion of the Earth. With translation, annotation and commentary and additional chapters on Ramus-Rheticus and the development of the problem before 1650* (Amsterdam, 1984).
Huiskamp, MW., P.J. Boon & Robert Louis Mathieu Maria Camps, *Catalogus van de pamfletten aanwezig in de Bibliotheek Arnhem 1537-1795* (Hilversum, 1995).
Hunt, Thomas Sterry, *Chemical and Geological Essays* (Boston, 1875).
Icones ad vivum delineatae et expressae, virorum clariorum qui praecipue scriptis Academiam Lugduno Batavorum illustrarunt (Leiden, 1609).
Ingegno, Alfonso, *Saggio sulla filosofia di Cardano* (Florence, 1980).
Jachaeus, Gilbertus [= Gilbert Jack], *Disputationum physicarum octava De elementis* ... [resp.] Iacobus Bontebal (Leiden, 1608).
Jacquart, Danielle, "Minima in Twelfth-century Medical Texts," in Lüthy, Murdoch & Newman, eds., *Late Medieval and Early Modern Corpuscular Matter Theories*, 39-56.
Jaeger, Frans Maurits, *Elementen en atomen eens en thans. Schetsen uit de ontwikkelingsgeschiedenis der elementenleer en atomistiek*, 2nd rev. ed. (Groningen, 1920).
——, "Over David van Goorle als atomist, en over het geslacht van Goorle in Noord-Nederland," *Oud-Holland* 36 (1918), 205-42.
Jaeger, Werner, "Das sogenannte IV. Buch der Meteorologie des Aristoteles," *Hermes*, 50 (1915), 118-36.
James I [King of England], *Translaet Vanden Brief des Conincks van groot Britaengien, etc. gheschreven anede Edele Grootmogende Heeren Staten-Generael vande Vereenichde Nederlanden. Aengaende de beroepinghe Conradi Vorstii* (s.l., 1611).

Javelli, Crisostomo, *Super tres libros Aristotelis de anima quaestiones subtilissimae, in quibus clarissime resolvuntur dubia Aristotelis et commentatoris, eaque ut plurimum decisa habentur iuxta Theomisticum dogma* (Venice, 1552).

Jensma, Goffe Theunis, "Inleiding," in Jensma et al., eds., *Universiteit te Franeker*, 11-39.

——, "Uit het huis van Arcerius. Acht 'artes'-studenten en hun opvattingen over wetenschap en maatschappij, 1589-1639," in Jensma et al., eds., *Universiteit te Franeker*, 452-69.

Jensma, Goffe Theunis, Franck R.H. Smit & Martin H.H. Engels, eds., *Universiteit te Franeker 1585-1811: Bijdragen tot de geschiedenis van de Friese Hogeschool* (Leeuwarden, 1985).

Joly, Bernard, *Descartes et la chimie* (Paris, 2011).

Judgement of the Synode Holden at Dort Concerning the Five Articles. As also their Sentence touching Conrad Vorstius (London, 1619).

Kangro, Hans, "Erklärungswert und Schwierigkeiten der Atomhypothese und ihrer Anwendung auf chemische Probleme in der ersten Hälfte des 17. Jahrhunderts," *Technikgeschichte* 35 (1968), 14-36.

Knuttel, Willem Pieter Cornelis, *Verboden Boeken in de Republiek der Vereenigde Nederlanden: beredeneerde catalogus* (The Hague, 1914).

——, *Catalogus van de pamfletten-verzameling berustende in de Koninklijke Bibliotheek*, improved ed., 9 pts in 10 vols. (Utrecht, 1978).

Kubbinga, Henricus Hermanus, *The Molecularization of the World Picture, or, The Rise of the Universum Arausiacum* (Groningen, 2009).

Kuhn, Thomas S., *The Essential Tension: Selected Studies in Scientific Tradition and Change* (Chicago, 1977).

Land, Jan Pieter Niolaas, *De wijsbegeerte in de Nederlanden* (Den Haag, 1899).

Langereis, Sandra, "De verzameling munten, gemmen en schelpen van Abraham Gorlaeus te Delft," in Ellinoor Bergvelt, Michiel Jonker & Agnes Wiechmann, eds., *Schatten in Delft. Burgers verzamelen 1600-1750* (Zwolle & Delft, 2002), 91-100.

Lasswitz, Kurd, *Geschichte der Atomistik vom Mittelalter bis Newton*. 2 vols. (Hamburg/Leipzig, 1890; repr. Hildesheim, 1963).

Leibniz, Gottfried Wilhelm, *Theodicy: Essays on the Goodness of God, the Freedom of Man and the Origin of Evil*, ed. Austin M. Farrer, transl. E.M. Huggard (New York, 2009).

Leijenhorst, Cees & Christoph Lüthy, "The Erosion of Aristotelianism. Confessional Physics in Early Modern Germany and the Dutch Republic," in Cees Leijenhorst, Christoph Lüthy & Johannes M.M.H. Thijssen, eds., *The Dynamics of Aristotelian Natural Philosophy from Antiquity to 1700* (Leiden, 2002), 375-411.

Leinsle, Ulrich Gottfried, *Das Ding und die Methode: Methodische Konstitution und Gegenstand der frühen protestantischen Metaphysik*, 2 vols. (Augsburg, 1985).
Libavius, Andreas, *Alchymia triumphans* (Frankfurt, 1607).
Lubbert, Sijbrand, *Declaratio responsionis Conradi Vorstii* (Franeker, 1611).
——, *Responsio ad pietatem Hugonis Grotii* (Franeker, 1614).
Lüthy, Christoph, "An Aristotelian Watchdog as Avant-Garde Physicist: Julius Caesar Scaliger," *The Monist*, 84 (2001), 542-61.
——, "Atoms and Corpuscles. Kurd Lasswitz and the Historiography of Atomism," *Intellectual News*, no. 11/12 (2003), 7-16.
——, "David Gorlaeus' Atomism, or: The Marriage of Protestant Metaphysics with Italian Natural Philosophy," in Lüthy, Murdoch & Newman, eds., *Late Medieval and Early Modern Corpuscular Matter Theories*, 245-90.
——, "Entia & Sphaerae. Due aspetti dell'atomismo bruniano," in Eugenio Canone, ed., *La filosofia di Giordano Bruno. Problemi ermeneutici et storiografici* (Florence, 2003), 165-98.
——, "The Invention of Atomist Iconography," in Wolfgang Lefèvre, Jürgen Renn & Urs Schoepflin, eds., *The Power of Images in Early Modern Science* (Basel, 2003), 117-38.
——, "Sennert's Slow Conversion from Hylemorphism to Atomism," *Graduate Faculty Philosophy Journal*, 26 (2005), 99-121.
——, Thoughts and Circumstances of Sébastien Basson. Analysis, Micro-History, Questions," *Early Science and Medicine*, 2 (1997), 1-73.
Lüthy, Christoph & Leen Spruit, "The Doctrine, Life, and Roman Trial of the Frisian Philosopher Henricus de Veno (1574-1613)," *Renaissance Quarterly*, 56 (2003), 1112-51.
Lüthy, Christoph, John E. Murdoch & William R. Newman, eds., *Late Medieval and Early Modern Corpuscular Matter Theories* (Leiden, 2001).
McKnight, Stephen A., "Francis Bacon's God," *The New Atlantis*, 10 (2005), 73-100.
Maclean, Ian, "The Interpretation of Natural Signs: Cardano's *De Subtilitate* vs. Scaliger's *Exercitationes*," in Brian Vickers, ed., *Occult and Scientific Mentalities in the Renaissance* (Cambridge, 1984), 231-52.
Mahnke, Dietrich, "Rektor Casmann in Stade. Ein vergessener Gegner aristotelischer Philosophie und Naturwissenschaft im 16. Jahrhundert," *Archiv für die Geschichte der Naturwissenschaften und der Technik*, 5 (1908), 183-97; 226-40, 352-63.
Maier, Anneliese, *Zwei Grundprobleme der scholastischen Naturphilosophie*, 2nd ed. (Rome, 1951).
Manzo, Silvia, *Entre el atomismo y la alquimia. La teoría de la materia de Francis Bacon* (Buenos Aires, 2006).

——, "Francis Bacon and Atomism: A Reappraisal," in Lüthy, Murdoch & Newman, eds., *Late Medieval and Early Modern Corpuscular Matter Theories*, 209-43.
Martijn, Caspar, *Christelijcke bekeeringhe, van Casper Martiin, te voren ghenaemt pater Sylvester van Carpentras, predicker van de ordre der cappucynen, en eerste overste van haer convent [...] in de ghereformeerde kercke van Oraengien op sondach den 21. december anno 1614* (Arnhem, 1615).
Mayer, H.C., "Ein Altdorfer Philosophenporträt," *Zeitschrift für Bayerische Kirchengeschichte*, 29 (1960), 145-66.
——, *Nikolaus Taurellus, der erste Philosoph im Luthertum. Ein Beitrag zum Problem Vernunft und Offenbarung* (Ph.D. theol., Göttingen, 1959).
Meinel, Christoph, "Early Seventeenth-Century Atomism. Theory, Epistemology, and the Insufficiency of Experiment," *Isis*, 79 (1988), 86-103.
——, "'Das letzte Blatt im Buch der Natur'. Die Wirklichkeit der Atome und die Antinomie der Anschuung in den Korpuskulartheorien der frühen Neuzeit," *Studia Leibnitiana*, 20 (1988), 1-18.
Menn, Stephen, "The Greatest Stumbling Block: Descartes' Denial of Real Qualities," in Roger Ariew & Marjorie Grene, eds., *Descartes and his Contemporaries. Meditations, Objections and Replies* (Chicago, 1995), 181-207.
——, "The Intellectual Setting," in *The Cambridge History of Seventeenth Century Philosophy*, eds. Daniel Garber & Michael Ayers (Cambridge, 1998), 33-86.
Mercenario, Arcangelo, *Dilucidationes obscuriorum locorum et quaestionum philosphiae naturalis* (Leipzig, 1585).
Mersenne, Marin, *L'impiété des déistes, athées et libertins de ce temps* (Paris, 1624).
——, *Quaestiones celeberrimae in Genesim* (Paris, 1623).
——, *La vérité des sciences contre les Sceptiques ou Pyrrhoniens* (Paris, 1625).
Meursius, Joannes, *Athenae Batavae, sive de urbe Leidensi et Academia, virisque claris qui utramque ingenio suo atque scriptis illustrarunt* (Leiden, 1625).
Michael, Emily, "Daniel Sennert on Matter and Form. At the Juncture of the Old and the New," *Early Science and Medicine*, 2 (1997), 272-99.
——, "Sennert's Sea Change: Atoms and Causes," in Lüthy, Murdoch & Newman, eds., *Late Medieval and Early Modern Corpuscular Matter Theories*, 331-62.
Morhof, Daniel Georg, *Polyhistor literarius, philosophicus et practicus*, with additions by Johannes Frick & Johannes Moller, 4th ed., 3 vols. (Lübeck, 1747).
Mühling, Andreas, "Arminius und die Herborner Theologen: Am Beispiel von Johannes Piscator," in Van Leeuwen et al., eds., *Arminius, Arminianism and Europe*, 115-34.
Muller, Richard A., "The Christological Problem in the Thought of Jacobus Arminius," *Nederlandsch Archief voor Kerkgeschiedenis*, 68 (1988), 145-63.
——, *God, Creation and Providence in the Thought of Jacob Arminius: Sources and Directions of Scholastic Protestantism in the Era of Early Orthodoxy* (Grand Rapids, 1991).

Mulsow, Martin & Jan Rohls, eds., *Socinianism and Arminianism: Antitrinitarians, Calvinists and Cultural Exchange in Seventeenth-Century Europe* (Leiden, 2005).
Murdison, Johannes, *Theses philosophicæ de primis rerum naturalium principiis...* [resp.] Hieronimus Belonius (Leiden, 1612).
Murdoch, John E., "The Medieval and Renaissance Tradition of *Minima Naturalia*," in Lüthy, Murdoch & Newman, eds., *Late Medieval and Early Modern Corpuscular Matter Theories*, 91-131.
Naudé, Gabriel, *Advis pour dresser une bibliothèque* (Paris, 1627).
Napius, W.J. & G.A. Lindeboom, *Hoogleraren in de Geneeskunde aan de Hogeschool en het Athenaeum te Franeker, 1585-1811* (Amsterdam, 1585).
Nauta, Lodi, "William of Ockham and Lorenzo Valla: False Friends. Semantics and Ontological Reduction," *Renaissance Quarterly*, 56 (2003), 613-51.
Nerdenus, Henricus Antonii, *Disputatio theologica de usura*. [Resp.] Henricus de Veno (Franeker, 1596).
Newman, William R., "The Alchemical Sources of Robert Boyle's Corpuscular Philosophy," *Annals of Science*, 53 (1996), 567-85.
——, *Atoms and Alchemy. Chymistry and the Experimental Origins of the Scientific Revolution* (Chicago, 2006).
——, "Experimental Corpuscular Theory in Aristotelian Alchemy: From Gebert to Sennert," in Lüthy, Murdoch & Newman, eds., *Late Medieval and Early Modern Corpuscular Matter Theories*, 291-329.
Nielsen, Lauge Olaf, "A Seventeenth-Century Physician on God and Atoms. Sebastian Basso," in Norman Kretzmann, ed., *Meaning and Inference in Medieval Philosophy. Studies in Memory of Jan Pinborg* (Dordrecht, 1988), 297-369.
Nienes, Arie Pieter van, J. Rinzema, M. Offringa, P.A.J. van den Berg et al., *De archieven van de Universiteit te Franeker, 1585-1812* (Leeuwarden, 1985).
Nieuw Nederlandsch Bibliografisch Woordenboek, eds. P.C. Molhuysen & P.J. Blok, 9 vols. (Leiden, 1911-1933).
Nifo, Agostino, *Subtilissima commentaria in libros Meteorologicorum, et in librum de mistis, sive quartum meteorum ab antiquis nuncupatum et ordinatum, longe fidelius et accuratius quam antea recognita et ab erroribus repurgata* (Venice, 1560).
Nijenhuis, Willem, "De Theologische Faculteit: Inleiding," in *Universiteit te Franeker*, ed. Jensma et al., 227-35.
Otterspeer, Willem, & Anneliese Aerts-van Bueren, *Werkplaatsen van vrijheid, geleerdheid en het ware geloof: of De wisselwerking tussen de universiteiten van Leiden en Franeker* (Franeker, 1985).
Partington, James Riddick, "The Origins of the Atomistic Theory," *Annals of Science*, 4 (1939), 245-82.
Palmerino, Carla Rita, "Bodies in Water like Planets in the Skies: Uses and Abuses of Analogical Reasoning in the Study of Planetary Motion," *Mechanics and*

Cosmology in the Medieval and Early Modern Period, eds. Massimo Bucciantini, Michele Camerota & Sophie Roux (Florence, 2007), 145-68.

——, "La fortuna della scienza galileiana nelle Provincie Unite," in L. Pepe & G.P. Brizzi, eds., *Galileo e la scuola galileiana nelle università del Seicento* (Bologna, 2011), 61-79.

Pasnau, Robert, *Metaphysical Themes 1274-1671* (Oxford, 2011).

Passmore, John, *Man's Responsibility over Nature* (London, 1974).

Paulian, Aimé-Henri, *Dictionnaire de physique, dédié au Roi*, 9th ed. (Nîmes, 1789).

Peeters, F.A.H., "Bij de heruitgave der Idea Physicae (1651) van David van Goorle, een onafhankelijk denker van nationale betekenis," *Bulletin van de Kring voor de Geschiedenis van de Pharmacie in Benelux*, 73 (1988), 28-36.

——, (ed.), *Davidis Gorlaei Ultrajectini, Idea Physicae (1651), met vertalingen in het Nederlands en Fries* (Tilburg, 1986).

Piccolomini, Francesco, *De mixtione*, in *Librorum ad scientiam de natura attinentium partes quinque*, 2 vols. (Venice, 1600), II: 98r-112v.

Pomponazzi, Pietro, *Dubitationes in quartum Meteorologicorum Aristotelis librum, nunc recens in lucem editae* (Venice, 1563).

Postma, Ferenc & Jan van Sluis, *Auditorium Academiae Franekerensis. Bibliographie der Reden, Disputationen und Gelegenheitsdruckwerke der Universität und des Athenäums in Franeker 1585-1743* (Leeuwarden, 1995).

Prosperi, Adriano, "Per la storia dell'Inquisitione romana," in Andrea Del Col & G. Paolin, eds, *L'inquisizione romana in Italia nell'età moderna: Archivi, problemi di metodo e nuove ricerche* (Rome, 1991), 27-64.

Ravensperger, Jacob, *Disputatio philosophica inauguralis: quam magnifici D. Rectoris D. Johannis Steinbergen … pro consequendo supremo in philosophia gradu sustinebit Jacobus Ravensberg* (Groningen, 1639).

Rees, Graham, "Atomism and 'Subtlety' in Francis Bacon's Philosophy," *Annals of Science*, 37 (1980), 549-71.

Regius, Henricus, *Physiologia, sive Cognitio sanitatis, tribus disputationibus in Academia Ultrajectina publice proposita* (Utrecht, 1641).

——, *Disputationes medicae de illustribus aliquot questionibus physiologicis* (Utrecht, 1641).

Reif, Patricia, *Natural Philosophy in Some Early Seventeenth-Century Scholastic Textbooks*, PhD thesis (St. Louis University, 1962).

——, "The Textbook Tradition in Natural Philosophy 1600-1650," *Journal for the History of Ideas*, 30 (1969), 17-32.

Reimmann, Jacob Friedrich, *Versuch einer Einleitung in die Historiam literariam insgemein und derer Teutschen insbesonderheit*, 3 vols. (Halle, 1710-1734).

Reneri, Henri, *Disputationum physicarum quarta, De elementis.* [Resp.] Paulus D. Keserui (Utrecht, 1635).

Requeste vande studenten in de H. theologie, ghelevert inde handen van … de heeren curateuren van de universiteyt tot Leyden …: Aengaende de beroepinghe Conradi Vorstij (Leiden, 1610).
Ribard, Dinah, *Raconter, vivre, penser: Histoires de philosophes (1650-1766)* (Paris, 2003).
Rodis-Lewis, Geneviève, "René Descartes," in Jean-Pierre Schobinger, ed., *Überweg Grundriss der Geschichte der Philosophie: Die Philosophie des 17. Jahrhunderts*, 4 parts in 8 vols. (Basel, 1998-2001), 2/1: 273-348.
——, "Problèmes discutés entre Descartes et Regius: L'âme et le corps," in Theo Verbeek, ed., *Descartes et Regius. Autour de l'explication de l'esprit humain* (Amsterdam, 1993), 35-46.
Rohls, Jan, "Calvinism, Arminianism and Socinianism in the Netherlands until the Synod of Dort," in Mulsow & Rohls, eds., *Socinianism and Arminianism*, 3-48.
Rother, Wolfgang, *Die Philosophie an der Universität Basel im 17. Jahrhundert. Quellen und Analyse* (PhD thesis, Universität Zürich, 1980).
Rothschuh, Karl, *Physiologie im Wandel ihrer Konzepte, Probleme und Methoden* (Freiburg, 1968).
Roux, Sophie, "An Empire Divided: French Natural Philosophy (1670-1690)," in *The Mechanization of Natural Philosophy*, eds. Daniel Garber and Sophie Roux (Berlin, 2012), [forthcoming]
Rübel, Rudolf, *Das Burgsteinfurter Gymnasium Arnoldinum im Wandel der Zeiten* (Burgsteinfurt, 1953).
Sabra, A.I., "The Simple Ontology of *Kalām* Atomism: An Outline," *Early Science and Medicine*, 14 (2009), 68-78.
Saguens, Johannes, *Systema Eucharisticum P. Maignani vindicatum ab impugnationibus contentis in opuscuolo dogmatico, quod scripit nuper R.P. Gennarus …* (Toulouse, 1705).
Sassen, Ferdinand, *Geschiedenis van de wijsbegeerte in Nederland tot het einde der negentiende eeuw* (Amsterdam, 1959).
——, "Grotius, philosophe aristotélicien," *Grotiana* 9 (1941/42), 37-52.
——, *Henricus Renerius, de eerste 'Cartesiaansche' hoogleraar te Utrecht* (Amsterdam, 1941).
Scaliger, Julius Caesar, *Exotericarum exercitationum liber XV. de subtilitate, ad Hieronymum Cardanum* (Paris, 1557; 2nd ed. Frankfurt, 1576).
Schmid, Xaver, *Nikolaus Taurellus, der erste deutsche Philosoph* (Erlangen, 1860).
Schoock, Martin, *Admiranda methodus novae philosophiae Renati des Cartes* (Utrecht, 1643).
Sennert, Daniel, *Epitome naturalis scientiae, comprehensa disputationibus viginti sex* (Wittenberg, 1600).

——, *De chymicorum cum Aristotelicis consensu ac disensu liber I* (Wittenberg, 1619, 2nd rev. ed., Wittenberg 1629).
——, *Hypomnemata physica* (Wittenberg, 1636).
Shackelford, Jole, "Seeds with a Mechanical Purpose: Severinus' Semina and Seventeenth-Century Matter Theory," in Allen G. Debus & Michael T. Walton, eds., *Reading the Bok of Nature: The Other Side of the Scientific Revolution* (Kirksville, 1998), 15-44.
Simoni, Anna E.C., ed., *Catalogue of Books from the Low Countries 1601-1621 in the British Library* (London, 1990).
Simonutti, Luisa, *Arminianesimo e tolleranza nel Seicento olandese: Il carteggio Ph. van Limborch – J. le Clerc* (Florence, 1984).
Skinner, Quentin, "Meaning and Understanding in the History of Ideas," *History and Theory*, 8 (1969), 3-53.
Slatius, Henricus [attr.], *Den gepredestineerden dief, ofte Een Samen-sprekinge ghehouden tusschen een predicant der Calvinus-ghesinde, ende een dief, die ghesententieert was om te sterven* (s.l., 1619).
Sorel, Charles, *De la perfection de l'homme, où les vrays biens sont considérez, et spécialement ceux de l'âme, avec les méthodes des sciences* (Paris, 1655).
——, *La Science des choses corporelles. Première partie de la science humaine, où l'on connoist la Verité de toutes les choses du Monde par les forces de la Raison et l'on trouve la refutation des Erreurs de la Philosophie vulgaire* (Paris, 1634).
Sozzini, Fausto, *De Sacrae Scripturae auctoritate libellus,* ed. Conrad Vorstius (Steinfurt, 1611).
——, *De officio hominis Christiani hominis in hodiernis istis de Religione controversiis* (Irenopolis, 1610 [=Franeker, 1610/11]).
——, *Het ampt van een christen mensch, door F. S. van Siena* [ed. Heinrich Welsing] (n.p., 1621).
Sprunger, Keith L., "William Ames and the Franeker Link to English and American Puritanism," in Jensma et al., eds., *Universiteit te Franeker*, 264-74.
Stanglin, Keith D., "Arminius and Arminianism: An Overview of Current Research," in Van Leeuwen et al., eds., *Arminius, Arminianism and Europe*, 3-24.
——, *Arminius on the Assurance of Salvation: The Context, Roots, and Shape of the Leiden Debate, 1603-1609* (Leiden, 2007).
Stellingwerff, Frederik, *De constitutione logices disputatio scholastica* (Franeker, 1610).
——, *Politycq discours: nopende den Staet van Frieslandt, waerinne bethoont wort, dat het selve Landt voor geen Slechte Heerlücheyt ghenomen voor een Leengoet ende minder als een Graefschap oyt heeft behooren geacht te worden ...* (Franeker, 1617).
Stolberg, Michael, "Particles of the Soul: The Medical and Lutheran Context of Daniel Sennert's Atomism," *Medicina nei Secoli*, 15 (2003), 177-224.

Sutton, Robert B., "The Phrase *Libertas philosophandi*," *Journal of the History of Ideas* 14 (1953), 310-16.
Taurellus, Nicolaus, *Arnaldi Villanoviani philosophi et medici summi opera omnia, cum … annotationibus* (Basel, 1585).
——, *Carmina funebria* (Nuremberg, 1602).
——, *Emblemata physico-ethica* (Nuremberg, 1602).
——, *KOSMOLOGIA, hoc est … De Mundo libri II* (Amberg, 1603).
——, *Medicae praedicationis methodus* (Frankfurt, 1581).
——, *De mundo et coelo libri IV, editio nova, cui accesserunt ejusdem autoris, 1. de usiis per se subsistentibus, 2. de communibus qualitatibus, tractatuli* (Amberg, 1611).
——, *Philosophiae triumphus* (Basel, 1573; 2nd ed., Arnhem, 1617).
——, *De rerum aeternitate, metaphysices universalis partes quatuor* (Marburg, 1604).
——, *Synopsis Aristotelis metaphysices ad normam Christianae religionis explicatae, emendate, et completae* (Hanau, 1596).
Tedeschi, John, *Il giudice e l'eretico. Studi sull'Inquisizione romana* (Milan, 1997).
Testimonium Academiae Ultrajectinae, et narratio historica quae defensae, qua exterminatae novae philosophiae (Utrecht, 1643).
Timpler, Clemens, *Disputatio metaphysica de signo et signato.* [Resp.] Augustinus Arnoldi, [Praes.] Henricus de Veno (Franeker, 1604). [Geneva, Bibliothèque Publique et Universitaire, Cd. 145-34].
——, *Metaphysicae systema methodicum, libris quinque comprehensum* (Steinfurt, 1604).
Toletus, Franciscus, *Commentaria una cum Quaestionibus, in octo libros Aristotelis de Physica auscultatione*, 2nd ed. (Venice, 1580).
Tolsma, Marijke, "*Iconographia Arminiana*: Portraits from 1609 until ca. 1850," in Van Leeuwen e.a., eds., *Arminius, Arminianism, and Europe*, 239-61.
Tremerius, Johannis & Antonius Fringilla, "Gorlaeus," *Chimica.* Leidsch chemisch periodiek, 1 (1971), 11.
Trutius, Antonius, *Theses logicae ex Categoriis; theses physicae de principiis.* [Resp.] Henricus de Veno (Leiden, 1593).
Twee disputatien van de Goddeliicke Predestinatie: d'eene by Doct. Franciscus Gomarus, d'ander by Doct. Iacobus Arminius (Leiden, 1609).
Valerius, Cornelius, *Physicae seu De naturae philosophia institutio* (Antwerp, 1584).
Valla, Lorenzo, *Repastinatio dialectice et philosophie*, ed. G. Zippel, 2 vols. (Padua, 1982).
Van Berkel, Klaas, "Franeker als centrum van ramisme," in Jensma et al., eds, *Universiteit te Franeker*, 424-37.
——, *Isaac Beeckman (1588-1637) en de mechanisering van het wereldbeeld* (Amsterdam, 1983).

——, *The Mechanical Philosophy in the Making* [forthcoming]
Van Buchell, Arend, *Diarium*, eds. G. Brom & L.A. van Langeraad (Amsterdam, 1907).
Van Bunge, Wiep, *From Stevin to Spinoza. An Essay on Philosophy in the Seventeenth-Century Dutch Republic* (Leiden, 2001).
Van den Broek, Jos M., *Van Albinusdreef tot Zeemanlaan: een wetenschappelijke wandeling langs Leidse straten, gebouwen, musea en instituten* (Leiden, 2005).
Van der Kemp, A.C.J., *Kerkelijk leven te Arnhem (1578-1815). Bijdrage tot de geschiedenis der Ned. Herv. Gemeente van Arnhem* (Arnhem, 1896).
Van der Woude, Cornelis, *Sibrandus Lubbertus. Leven en werken, in het bijzonder naar zijn correspondentie* (PhD thesis, Vrije Universiteit Amsterdam, 1963).
Van Deursen, Arie Theodorus, "The Dutch Republic, 1588-1780," in Blom & Lamberts, eds., *History of the Low Countries*, 142-218.
Van Gelder, H.A. Enno, *Getemperde vrijheid. Een verhandeling over de verhoudingen van Kerk en Staat in de Republiek der Verenigde Nederlanden en de vrijheid van meningsuiting in zake godsdienst, drukpers en onderwijs, gedurende de 17e eeuw* (Groningen, 1972).
Van Goorle, Peter, *Prognosticatie vanden Jare ons Heeren MCCCCCLXV ghecalculeert opter elevatien van Antwerpen* (Antwerp, 1564).
Van Huisstede, Pieter & J.P.J. Brandhorst, *Dutch Printer's Devices 15th-17th Century* (Nieuwkoop, 1999).
Van Kleef, Frans, "Redactioneel," in *Chimica*, 1/1 (1970), 1-2.
Van Leeuwen, Th. Marius, "Introduction: Arminius, Arminianism, and Europe," in Van Leeuwen et al., eds., *Arminius, Arminianism, and Europe*, ix-xii.
—— "De erfenis van Arminius," in Van Holtzapffel et al., *De Remonstrantie 400 jaar*, 23-33.
Van Leeuwen, Th. Marius, Keith D. Stanglin & Marijke Tolsma, eds., *Arminius, Arminianism, and Europe. Jacob Arminius (1559/60-1609)* (Leiden, 2009).
Van Limborch, Philippe, ed., *Praestantium ac eruditorum virorum epistolae ecclesiasticae et theologicae*, 3rd rev. ed. (Amsterdam, 1704).
Van Melsen, Andreas G.M., *Van Atomos naar Atoom. De geschiedenis van het begrip Atoom* (Amsterdam, 1949).
——, *From Atomos to Atom. The History of the Concept of Atom*, trans. H.J. Koren (Pittsburgh, 1952).
Van Nienes, Arie Pieter, Jacob Rinzema, M. Offringa & Peter Arend Jan van den Berg, eds., *De archieven van de Universiteit te Franeker, 1585-1812* (Leeuwarden, 1985).
Van Nieuwenburg, Cornelis Johannes, *Korte geschiedenis van de chemie* (Hilversum, 1961).
Van Ruler, Johannes A., *The Crisis of Causality. Voetius and Descartes on God, Nature and Change* (Leiden, 1995).

Van Slee, Jacob Cornelis, *De geschiedenis van het Socinianisme in de Nederlanden* (Haarlem, 1914).

Van Spaendonck, Hein A., *Catalogus van de Arnhemse drukken tot 1800 aanwezig in de Bibliotheek Arnhem* (Hilversum, 2002).

Van Swanenburg, Willem, *Icones ad vivum delineatae et expressae, virorum clariorum qui praecipue scriptis Academiam Lugduno Batavorum illustrarunt* (Leiden, 1609).

Verbeek, Theo, *Descartes and the Dutch. Early Reactions to Cartesian Philosophy, 1637-1650* (Carbondale and Edwardsville, 1992).

——, "*Ens per accidens.* Le origini della *Querelle* di Utrecht," *Giornale critico della filosofia italiana*, 6th series, 71 (1992), 276-88.

——, "Gassendi et les Pays Bas," in Sylvia Murr, ed., *Gassendi et l'Europe (1592-1792). Actes du Colloque international de Paris 'Gassendi et sa postérité (1592-1792)'. Sorbonne, 6-10 octobre 1992* (Paris, 1997), 263-73.

——, *René Descartes et Martin Schoock: La Querelle d'Utrecht* (Paris, 1988).

——, ed., *Descartes et Regius. Autour de l'Explication de l'esprit humain* (Amsterdam, 1993).

Verhel, Arnold, *Speculum philosophiae primae entis* (Franeker, 1662).

Vermij, Rienk, *The Calvinist Copernicans. The Reception of the New Astronomy in the Dutch Republic, 1575-1750* (Amsterdam, 2002).

Voetius, Gijsbert, *Catechisatie over den Catechismus der Remonstranten* … (Utrecht, 1641).

——, *Selectarum disputationum fasciculus*, ed. D. Abraham Kuyper (Amsterdam, 1886).

——, *Selectarum disputationum theologicarum pars prima* (Utrecht, 1648).

Voetius, Paulus, *Prima philosophia reformata* (Utrecht, 1657).

Vorstius, Aelius Averhardus, *Theses physicæ de Elementis* … [resp.] Cornelius Petri F. Burchuliet (Leiden, 1609).

Vorstius, Conrad, *Anti-Bellarminus contractus: hoc est, compendiosum examen omnium fidei controversiarum* (Hanau, 1610).

——, *Christiana et modesta responsio ad articulos quosdam* … (Leiden, 1611).

——, *Tractatus theologicus de Deo, sive De natura & attributis Dei* (Steinfurt, 1610).

Vriemoet, Emon Lucius, *Athenarum Frisiacarum libri II* (Franeker, 1758).

Wackernagel, Hans Georg et al., eds., *Die Matrikel der Universität Basel.* 5 vols. (Basel, 1951-1980).

Weidler, Johannes Friedrich, *Historia astronomiae, sive De ortu et progressu astronomiae* (Wittenberg, 1741).

Wenneker, Erich, "Vorstius, Conrad," in *Biographisch-Bibliographisch Kirchenlexikon*, vol. 13 (Herzberg, 1998), col. 84-90.

Westman, Robert S., *The Copernican Question. Prognostication, Scepticism, and Celestial Order* (Berkeley, 2011).

Whyte, Lancelot Law, *Essay on Atomism: from Democritus to 1960* (London, 1961).
Wierda, Gerben, "Twee studiegenoten van David. Over Frederic Stellingwerff en Engelbertus Aegidius" (unpublished article).
Winer, Georg Benedikt, *Comparative Darstellung des Lehrbegriffs der verschiedenen christlichen Kirchenparteien, nebst vollständigen Belegen aus den symbolischen Schriften derselben*, 2nd ed. (Leipzig, 1837).
Wtenbogaert, Johannes, *Tractaet Van t'ampt ende authoriteyt eener hoogher christelicker overheydt, in kerckelicke saecken* (The Hague, 1610).
Wybinga, Martinus à, *Laudatio funebris habita post exequias in templo academico in honorem ac memoriam … D. Pierii Winsemii …* (Leeuwarden, 1645).
Zanchi, Geronimo, *De natura Dei sive De divinis attributis libri V* (Heidelberg, 1577).
Zedler, Johann Heinrich, *Grosses vollständiges Universal-Lexikon aller Wissenschaften und Künste*, 68 vols. (Leipzig & Halle, 1732-1754; repr. Graz, 1961-64).
Zeltner, Gustav Georg, *Historia crypto-Socininismi, Altorfinae quondam Academiae infesti* (Leipzig, 1729).
Zijlstra, Samme, *Het geleerde Friesland – een mythe? Universiteit en maatschappij in Friesland en Stad en Lande ca. 1380-1650* (Leeuwarden, 1996).
Zubov, Vasilij Pavlovič [= Walter Subow], "Zur Geschichte des Kampfes zwischen dem Atomismus und dem Aristotelismus im 17. Jahrhundert (*Minima naturalia* und *Mixtio*)," in Gerhard Harig, ed., *Sowjetische Beiträge zur Geschichte der Wissenschaft* (Berlin, 1960), 161-91.

Index

E

F

G

H

I

J

K

L

M

S

T

V

W

Z